Soy Protein-Based Blends, Composites and Nanocomposites

Scrivener Publishing
100 Cummings Center, Suite 541J
Beverly, MA 01915-6106

Publishers at Scrivener
Martin Scrivener (martin@scrivenerpublishing.com)
Phillip Carmical (pcarmical@scrivenerpublishing.com)

Soy Protein-Based Blends, Composites and Nanocomposites

Edited by

Visakh P. M and Olga Nazarenko

Scrivener Publishing

WILEY

This edition first published 2017 by John Wiley & Sons, Inc., 111 River Street, Hoboken, NJ 07030, USA and Scrivener Publishing LLC, 100 Cummings Center, Suite 541J, Beverly, MA 01915, USA

For more information about Scrivener publications please visit www.scrivenerpublishing.com.

Wiley Global Headquarters
111 River Street, Hoboken, NJ 07030, USA

For details of our global editorial offices, customer services, and more information about Wiley products visit us at www.wiley.com.

Library of Congress Cataloging-in-Publication Data
ISBN 978-1-119-41830-6

Cover image: Pixabay.Com
Cover design by Russell Richardson

Set in size of 11pt and Minion Pro by Exeter Premedia Services Private Ltd., Chennai, India

10 9 8 7 6 5 4 3 2 1

Contents

Preface

Many of the recent research accomplishments in the area of soy-based blends, composites and bionanocomposites are presented in this book. In addition to introducing soy protein and its structure and relationship properties, an attempt has been made to cover many other relevant topics such as the state-of-the-art, new challenges, advances and opportunities in the field; biomedical applications of soy protein; electrospinning of soy protein nanofibers, their synthesis and applications; soy proteins as a potential source of active peptides of nutraceutical significance; soy protein isolate-based films; and use of soy protein-based carriers for encapsulating bioactive ingredients.

This book is intended to serve as a one-stop reference resource for important research accomplishments in the area of soy protein-based biocomposites and bionanocomposites. It will be a very valuable reference source for university and college faculties, professionals, post-doctoral research fellows, senior graduate students, and researchers from R&D laboratories working in the area of soy protein and its biocomposites and bionanocomposites. Since the various chapters in this book have been contributed by prominent researchers from industry, academia and government/private research laboratories across the globe, it is an up-to-date record of the major findings and observations in the field. The first chapter acts as an introduction to soy protein-based blends, composites and nanocomposites, including their scope, state of the art, preparation methods, environmental concerns regarding nanoparticles and related challenges and opportunities.

Included in the second chapter introducing general aspects of soy proteins, is a discussion of their source, structure and relationship properties. Chemical modification and characterization of soy proteins are also included in this chapter along with a description of the applications of soy protein-based nanocomposites and blends. Advances in soy protein-based nanocomposites are addressed in the third chapter, in which the authors discuss how the incorporation of nanoparticles proves to be an effective way to improve physical properties, especially the mechanical properties and water resistance which limit their extensive use. The properties of the resulting nanocomposites are highly dependent on the processing methods, nature of nanofillers, as well as the dispersion effect of the filler in the matrix. Therefore, the fabrication methods, property-structure relationship, and application of soy protein nanocomposites are also reviewed in this chapter.

The following chapter on applications of soy protein-based blends, composites and nanocomposites discusses many topics, including the particulars of soy protein applications, soy protein-based blends, and those of soy protein-based nanocomposites. The fifth chapter based on biomedical applications of soy protein summarizes many of the recent accomplishments in the area of biomedical research. In this chapter, the authors discuss various topics such as forms and properties of soy proteins, application of plant protein in biomedical applications, application of soy proteins in wound dressings, and the potential use of soy proteins in products and applications in regenerative medicine, tissue engineering, and drug delivery systems.

The following chapter is a good structural basis for the understanding of electrospinning of soy protein nanofibers. Discussed in the chapter are the production of nanofibers from different synthetic and natural polymers, the physical properties of soy proteins that affect their electrospinning, followed by a summary of relevant work that has been done in the area. The chapter closes with a discussion on possible applications of electrospun nanofibers from soy proteins. The use of soy proteins as a potential source of active peptides of nutraceutical significance is the subject of the seventh chapter, which introduces the main concepts along with examples to help readers understand them. This chapter is devoted to reviewing the literature to identify and describe the available methodologies for the identification and production of bioactive peptides from soybean proteins. In addition, potential applications of these peptides as functional foods and therapeutic agents are also highlighted.

The authors of the eighth chapter present a brief account of the topic of soy protein isolate-based films, including soy protein film preparation, characterization of soy protein films, modifications and applications. The last chapter of the book reviews recent progress in the preparation of soy protein-based carriers for bioactive ingredients encapsulation.

In conclusion, the editors would like to express their sincere gratitude to all of the contributors to this book for their excellent support in the successful completion of this venture. We are grateful to them for the commitment and sincerity they have shown towards their contributions. Without their enthusiasm and support, this book would not have been possible.

We would also like to thank all the reviewers who have taken their valuable time to make critical comments on each chapter. We also thank the publisher John Wiley and Sons Ltd. and Scrivener Publishing for recognizing the demand for such a book, realizing the increasing importance of the area of soy protein-based blends, composites and nanocomposites, and for starting such a new project, which not many other publishers have handled.

Visakh. P. M
Olga Nazarenko
Tomsk, Russia
June 2017

1

Soy Protein: State-of-the-Art, New Challenges and Opportunities

Visakh P. M

Department of Ecology and Basic Safety, Tomsk Polytechnic University, Tomsk, Russia

Abstract
This chapter deals with a brief account on various topics in rubber-based bionanocomposites: Preparation and state-of-the-art. It also discusses different topics such as soy protein: Introduction, structure and properties relationship, thermoplastic-based soy protein nanocomposites, applications of soy protein-based blends, composites, and nanocomposites, biomedical application of soy protein, preparation of soy protein nanofibers by electrospinning, physiologically active peptides derived from soy protein, soy protein polymer-based (film) membranes and encapsulation of bio actives using soy protein-based material.

***Keywords*:** Rubber-based bionanocomposites, soy protein, soy protein nanocomposites, soy protein nanofibers

1.1 Soy Protein: Introduction, Structure and Properties Relationship

Soy proteins are one of the most abundant and most widely utilized plant proteins on this planet. With high content of essential amino acid and desirable functional properties, soy proteins have attracted persisting interest in food and pharmaceutical industry. The 11S and 7S globulins represent approximately 60% of the storage protein in soybeans. They are the most important contributors to the physicochemical and functional properties of

Corresponding author: visagam143@gmail.com

Visakh P. M and Olga Nazarenko (eds.) Soy Protein-Based Blends, Composites and Nanocomposites, (1–22) 2017 soy protein products. It exhibits a high content of negatively charged amino acids such as glutamic acid and aspartic acid, whereas the percentage of hydrophobic amino acids such as leucine is relatively low [1]. β-conglycinin is relatively flexible, as evidenced by its high contents of α-helix, and random coils [2]. It comprises six major isomers, each of which is composed of three major subunits and two minor ones (γ and δ) [3].

As can be seen, the whole soybean seed is cleaned, cracked, dehulled, and flaked to produce soy powder. The powder is then subjected to oil extraction with organic solvents such as hexane. The particle sizes range from grits (or flakes) of varying sieve specification to fine powders. The soy meal could be further ground into soy flour (SF), a product that contains less than 1% oil and a protein content ranging from 40–60%. Soy proteins with higher purity may also be produced with smaller particles, because the protein can be more effectively extracted from finer flours, making the separation of protein from insoluble carbohydrate more efficient and complete.

There exist three major types of soy protein-rich products, SPC (soy protein concentrate, and fractionated 11S/7S globulins (protein content >90%, fraction purity >85%). Quite a few methods have been developed to produce these products with desirable features, and several typical approaches with respect to their principles, major procedures, advantages, and drawbacks. The majority of the protein is precipitated and recovered by a second centrifugation. The curd-like precipitate is neutralized with alkali, washed with water to remove excessive alkali and salt, and finally spray dried or lyophilized to yield the final product. The typical protein of yield (weight ratio between the product and the raw material) is around 30%, though a yield of as high as 44% has been reported [4]. The product loses part of its original solubility as a result, but it gains some desirable properties such as good texture and water holding capacity [5].

In pilot scale production, the solvent such as alcohol and hot water can be recovered through evaporation and condensation, thus achieving higher extraction efficiency. In addition to the traditional methods, membrane-based techniques including micro- and ultrafiltration have also been widely studied for the preparation of SPC. Teng *et al.* further investigated the effect of divalent cations on the fractionation process. They suggested that using Mg^{2+} instead of Ca^{2+} as a precipitant improved the purities of both fractions without affecting their yields significantly. Soy proteins tend to adopt a compactly folded structure, with their hydrophilic and charged amino acid residues maximally exposed to the solvent and hydrophobic moieties buried in the globular core. The surface charge of colloidal particles is usually gauged by the electrical potential at the interfacial double layer at the location of the slipping plane relative to a point in the bulk fluid away from the interface. Proteins as amphiphilic molecules bear both hydrophilic and hydrophobic groups which endow their ability to interact with both the polar and nonpolar solvents and serve as an emulsifier [6].

Two parameters are commonly referred to when describing the emulsifying properties of a molecule. As many other proteins, soy proteins show viscoelasticity when dispersed in water. Under room temperature without the addition of cross-linkers (such as transglutaminase or calcium salts), the dispersion exhibits viscous property (G″) as the predominant characteristic [7]. Since viscosity is indicative for the friction between the molecule and the solvent, it is highly dependent on the interaction between them. Heated soy protein films exhibit decreased water vapor permeability, and increased percentage of elongation at break (%E) when compared to unheated ones [8].

While thermal denaturation is conventionally considered as a detrimental factor for protein solubility, combination of thermal treatment with a suitable pressure may make the protein more soluble. Glycerol is by now the most widely utilized plasticizer for soy protein-based plastics, owing to its relatively short and flexible chain as well as its strong hydrophilicity. The former character facilitates the insertion of glycerol into the peptide chains in the soy proteins, and the latter one promotes its interaction with the protein via extensive hydrogen bonding. Soy proteins are rich in both amine and carboxyl groups; therefore, they can readily react with additional carboxyl or amine groups. The reaction between the positively charged amine groups on the soy proteins and an external carboxylic acid is comparable to phosphorylation.

1.2 Advances in Soy Protein-Based Nanocomposites

Residual soy proteins, a by-product of the soy oil industry, are currently utilized in applications such as animal feed and food supplement. Soy proteins are composed of a mixture of albumins and globulins, 90% of which are storage proteins with globular structure, consisting mainly 7S (conglycinin) and 11S (glycinin) globulins. Soy protein contains 18 amino acids including those containing polar functional groups, such as carboxyl, amine, and hydroxyl groups that are capable of chemically reacting and making soy protein easily modification [9]. Biopolymer films are usually plasticized by hydroxyl compounds [10]. Glycerol has a high boiling point and good stability, and is regarded as one of the most efficient plasticizers for soy protein plastics [11]. Glycerol-plasticized soy protein possesses good processing properties and mechanical performance [12]. The bio-nanocomposites consist of a biopolymer matrix reinforced with particles having at least one dimension in the nanometer range (1–100 nm) and exhibit much improved properties due to high aspect ratio and high surface area of the nanoparticles.

Soy protein films reinforced with starch nanocrystals (SNC) could be prepared by casting method [13]. The SNC synthesis was developed by acid hydrolysis of native cornstarch. Soy protein is one of the few natural polymers that can be thermoplastically processed under the plasticization of small molecules [14]. Soy protein plastics without any additive have a brittle behavior, which makes processing difficult. Addition of plasticizers is an effective way to improve the flowability of soy protein melts and obtain flexible soy protein-based films. Phthalic anhydride modified soy protein (PAS)/glycerol plasticized soy protein (GPS) composite films were fabricated by using extrusion and compression-moulding [15]. Soy/BN nanocomposites were prepared by low-cost green technique with water as the solvent. The thermal properties of the nanocomposites were studied by thermogravimetric analysis (TGA). The biodegradation behaviors of maleated PCL/ isolated soy protein (SPI) composites reinforced with organoclay were evaluated by soil burial test [16]. Composites containing higher percentage of soy protein degraded rapidly in the initial 8 weeks and a gradual decrease of weight occurred during the next 8 weeks.

Soy protein films are effective barriers to the passage of lipid, oxygen, and carbon dioxide. However, the inherent hydrophilicity of proteins and the substantial amount of plasticizer added in the film perform poorly in moisture barrier and mechanical properties as packaging material. In addition of *in situ* synthesis, soy protein/silica nanocomposites could be fabricated through compounding nano-SiO_2 particles into soy protein isolate matrix [17]. Zheng *et al.* reported the nanocomposite sheets by compounding MWNTs of various sizes into SPI matrix through solution mixing and then compression-molding method [18]. Blending SPI with other biodegradable polymers such as polycaprolactone, poly(lactic acid), poly(vinyl alcohol), natural rubber, etc., thus becomes a way to enlarge its applications. The properties of the blend materials could further improved by nanoreinforcing. Sasmal *et al.* prepared a kind of bio-based, eco-friendly nanocomposites from maleated polycaprolactone/soy protein isolate blend (50/50 wt/wt) reinforced with organo-modified clay by melt compounding [19]. Soy protein plastics possess good mechanical strength and water resistance by compositing PS nanoparticles into soy protein matrix. The water uptakes of the nanoblends ranged from 11% to 19%, which is much lower than that of pure SPI (32%) at 75% RH.

1.3 Applications of Soy Protein-Based Blends, Composites, and Nanocomposites

Soybean is an important raw material for food industry, because is one of the most consumed grains in the world. Special applications of soy protein for development of biomaterials, composites, nanocomposites, and blends have been reported as potential use of this grain in several areas, such as biomedical, optoelectronic, optical coating, and packaging [20–24]. The soy protein nanocomposites can be used as adhesives, asphalt, resins, cleaning products, paper coatings, plastics, polyesters, and textile fibers that allow its use over a large area, such as packaging, medical, aerospace, and automotive fields. Meikle *et al.* [25] studied soybean-based hydrogels with different physicochemical properties and bioactivity, which were obtained by sequential or simultaneous procedure of SF defatting process and material extraction. SPI the major component of soybean has been used to prepare biodegradable materials, such as adhesives, plastics, and various binders in recent years.

Although the SPI plastics usually possess good biodegradability, their application is limited by poor flexibility and water resistance. Wang *et al.* [26] in which they incorporate cellulose whiskers to SPI aiming to improve mechanical properties, the authors obtained composites which showed greater water resistance and thermal stability. The improvement in the properties of the SPI/cellulose whisker composites may be ascribed to cross-linking networks caused by intermolecular hydrogen bonds between the cellulose whiskers and the SPI matrix. SPI is a biopolymer that has potential applications in packaging (films or coatings), because it offers interesting film-forming properties, good barrier properties to oxygen, aromas and lipids when in low to intermediate moisture conditions, besides it is a low-cost raw material [27].

The behavior of soy protein packaging systems is the modification of SPI with cross-linking agents. Cross-linking agents can work improving the mechanical and water barrier properties of soy protein films reducing its solubility, ability to swell, and gas/water vapor permeability. The SPI cross-linked with genipin was considered promising natural biodegradable materials for use in food packaging. Soy protein has many unique properties such as low cost, ease of handling, low press temperatures, and the ability to bind wood with relatively high moisture content, representing a very practical and inexpensive material for wood adhesives [28]. Thames *et al.* [29] developed a water resistant soy protein-based adhesive blend with polyol plasticizer, preferred glycerol, and with a vegetable oil derivative, aleinized methyl ester of tung oil. This adhesive can be useful in the manufacture of particleboard and other composites. SPI films high water vapor permeability is the most important obstacle to its use in food coatings [30], drawback which can be exceeded by blending methodology and thus many works have been developed in this area by blending.

Luo *et al.* [31] prepared a series of cellulose/SPI membranes, and observed that porous structure and the size of the pores in the surfaces increased with an increase of SPI content, and the incorporation of SPI in cellulose changed the compositions and microstructure, improving the biocompatibility of the membranes. In addition, the application of the PLA layer presented an important effect on the mechanical properties of the films, decreasing the elongation at break and increasing the tensile strength and the Young's modulus, resulting in a material less elongable and more resistant compared to those of pure SPI films. The resultant film made of soy/MMT is recommended to be used for packing food with high moisture content as fresh fruit and vegetable in order to replace low-density polyethylene (LDPE) and polyvinylidene chloride (PVDC).

1.4 Biomedical Applications of Soy Protein

Soybeans also contain a high amount of phytic acid that is antioxidant and can inhibit the growth of cancer cells, reduces blood sugar level and inflammation [32–34]. They are also a good source of fibre, iron, calcium, zinc, and vitamin B [35]. Soy Protein products can swell when they absorb water or can dissolve in water and this is an important functional property in drug delivery systems. In a research study by Ramnath *et al.*, composite biomaterials prepared from SP and sago starch cross-linked with gluteraldehyde were prepared as temporary wound-dressing materials [36]. Chein *et al.* studied the biocompatibility of SP scaffolds fabricated by freeze-drying and three-dimensional printing [37].

The content of SP in the scaffolds was varied. It was assessed using a subcutaneous implant model in female BALB/c mice age 6–8 weeks. These results indicated that SP is a potential biocompatible implant for tissue regeneration. The scaffold porosity, soy protein density, and scaffold degradation rate significantly affected the acute and humoral immune response. Chien and Shah prepared porous SP-based scaffolds [38]. Xu *et al.* reported the preparation of water-stable electrospun SP-based scaffolds [39]. The scaffolds had large volume and ultrafine fibres oriented randomly and evenly in three dimensions. They were used to simulate native extracellular matrices of soft tissues. In another research report, the parameters for electrospinning fibrous scaffolds from SP isolate by the addition of poly(ethylene oxide) dissolved in 1,1,3,3,3-hexafluoro-2-propanal were investigated. Their physicochemical properties were studied and they were found to exhibit mechanical properties that are similar to human skin. Silva *et al.* reported soy- and casein-based membranes for biomedical applications [40]. The membranes were subjected to cross-linking with glyoxal and tannic acid followed by thermal treatment. The cytotoxicity of both soy- and casein-based protein biomaterials were evaluated and it correlated with the materials degradation behavior. The SP isolate/poly(ethylene oxide) mats were cross-linked using carbodiimide to increase its robustness. SP isolate/poly(ethylene oxide) fiber diameters ranged between 50 nm and 270 nm depending on electrospinning and solution parameters. Soy hydrogels were injected into the subcutaneous pocket of mice and histological staining showed minimal fibrous capsule formation up to 20 days. It was found to be a potential biomaterial for tissue engineering and drug delivery applications [41].

A self-hardening soy/gelatine/hydroxyapatite composite foam was prepared and it was able to retain porosity upon injection. The foamed paste produced a calcium-deficient hydroxyapatite scaffold after setting. Implantation of the soybeans biomaterial over a period of 8 weeks produced bone repair with features distinct from those obtained by healing in a nontreated defect. New and progressively maturing trabeculae appeared in the animal group where soybeans biomaterial granules were implanted whereas; the sham operation produced only a rim of pseudo-cortical bone still featuring a large defect. Chitosan and soybean protein isolate blended membranes were prepared by solvent casting. These membranes exhibited a biphasic structure that originates *in situ* porous formation, through a two-step degradation mechanism. Vaz *et al.* reported SP drug delivery matrix systems prepared by melt-processing techniques, namely extrusion and injection moulding [42]. The soy matrix systems were encapsulated with theophylline drug by extrusion and cross-linking with glyoxal. Reddy *et al.* demonstrated the potential of SP isolate films as a drug release system for naturally occurring antiproliferative agent [43]. The films were prepared by casting method and the percentage of the resorcinol was varied between 10% and 30%.

In a research study by Chien *et al.*, SP hydrogels were developed by varying the weight percentages of water (15 wt.%, 18 wt.%, and 20 wt.%) [44]. Chemical modifiers or cross-linkers were not used to prepare the hydrogels. This method was useful for developing hydrogels for direct injection *in vivo*. The concentration of SP was varied and it influenced the rheological, swelling, mechanical properties and the release of the model drug, fluorescein from the hydrogels *in vitro*.

1.5 Electrospinning of Soy Protein Nanofibers: Synthesis and Applications

These protein-rich products have found uses in many non-food industrial applications, including the manufacture of plastics, adhesives, paper coatings, paint coatings, and composites. The advantages of using soy proteins in such systems are not only due to their physicochemical properties but also due to their renewability and sustainable production. Soybean proteins alone or in combination with other natural and synthetic polymers have been used to produce nanofibers by the electrospinning technique. In the solution electrospinning process, a polymer is first dissolved in a given solvent and the solution is pumped through a nozzle that together with a metallic fiber collector serve as the electrodes between which an electric field is applied.

The storage protein in soybeans accounts for a large fraction of the raw bean weight (between 65% and 80%) [45]. Storage proteins are globulins, that is, their solubility in water is enhanced by the presence of electrolytes. They have been classified according to the sedimentation constant as 7S and 11S or β conglycinin and glycinin, respectively. Reducing agents such as 2-mercaptoethanol, cysteine, NaCN, and dithiothreitol (DTT) have been used to break disulfide bonds in soy protein. Glycinin contains 2 free mol of sulfhydryl group/mol protein in its native state and 2–3 mol of sulfhydryl/mol after heating [46].

As soy proteins are globulins, their solubility in water is enhanced by the content of electrolytes. In addition, the pH will affect the solubility. The isoelectric pH of soy proteins has been reported as 4.5. The solubility will be low to zero at pH values near the isoelectric pH and increased at higher pH values. The solvents included water, acetic acid, ethanol, hydrochloric acid, acetone, sodium hydroxide, ammonium hydroxide, and some polar but less water soluble solvents, namely dimethylformamide (DMF), tetrahydrofuran (THF), and 1,1,1,3,3,3 Hexafluoro-2-propanol (HFIP). Pure soy protein globulins are rarely used in practical applications due to the cost involved in purification. Furthermore, the generalization is not straight forward, and the selection of a "suitable" solvent must fulfill the requirements of dissolving the solute within a reasonable time, being good solvent and environmentally friendly as well. The use of pure protein fractions could be also of little practical interest. That is why most reports have focused on the use of commercially available soy protein. Typically isolate has been used because of its high protein content; however, its cost is still high compared with SF which has been used very recently [47, 48] on electrospinning applications.

Additional complications in term of solubility and preparation of solutions for electrospinning arise when soy protein is blended with other polymers. This is because the solvent of choice has to be a good solvent for both the protein and the coadjutant polymer. This implies that the coadjutant polymer should be water soluble. To this end, polyethylene oxide and polyvinyl alcohol have been used. Other biopolymers include corn zein, wheat protein (gluten), and lignin [49]. The fibers were uniform, the blends with zein reduced drastically the content of soy protein on the final fibers, that is, the fibers were zein fibers containing soy protein [50].

A follow-up study of the same system [51] included the effect of changing the pH (9 and 12) on mechanical properties and biodegradability of soy protein isolate-polyvinyl alcohol elctrospun fibers. The nanofiber mats prepared from solutions at pH 9 exhibited a higher load and elongation at break than those prepared at pH 12. This effect was ascribed to the lower denaturation of the proteins at pH 9 compared with pH 12. The nanofiber samples exhibited low contact angles (high wettability), which could limit their practical application [52]. The synergy between soy proteins with other natural polymers to produce electrospun nanofibers has remained essentially unexplored, with only blends of zein-soy protein isolate-SF-gluten and soy protein isolate-lignin being reported.

Because of its high protein content, soy protein isolate has been the product of preference in most electrospinning of soy protein reports; however, the use of SF was reported recently. Soy protein/PVA (9 kDa and 130 kDa) and soy protein/PCL (80 kDa) fibers were electrospun on top of a rayon support membrane. Blends of SPI-PEO (1:4, 2:3, and 3:2) were used to prepare electrospun nanofibers and tested for wound-dressing applications [53]. SPI in dilute NaOH and PEO (300 kDa) in ethanol was prepared separately, then blended at different ratios of SPI: PEO before electrospinning.

The SPI/PEO mats exhibit antibacterial activity against one gram-negative (*Pseudomonas aeruginosa*) and one gram-positive bacteria (*Staphylococcus aureus*) as determined by the disk diffusion method. Although the experiments for wound-healing effect were qualitative observations on winstar rats (i.e., no quantification of ephitelial cell growth), the comparison indicates that the wounds covered with SPI/PEO electrospun mats exhibit a slightly better healing ability compared to uncovered wound. The produced soy protein scaffolds (3 cm × 3 cm) were applied to the wounds at different times. The results indicate that the protein scaffolds in contact with the skin get completely hydrated and dissolved in the wound. In addition the bacterial filtration efficiency increased with the load of nanofibers, with 5 g/m^2 exhibiting the highest efficiency. It was hypothesized that the adhesion of bacteria to the nanofibers might be as a result of electrostatic interactions, owed to the charge balance of amino-acids on the protein.

1.6 Soy Proteins as Potential Source of Active Peptides of Nutraceutical Significance

Recent developments in food bioactive protein/peptide databases, coupled with improved knowledge of various enzyme specificities can be used in a process known as *in silico* hydrolysis for the identification of potential bioactive peptides from food/soy proteins. *In silico* produced peptides with known sequences can then be subjected to quantitative structure–activity relationship (QSAR) studies for a preliminary assessment of their bioactivity potential. Fermentation of soybean has been shown to result in the release of peptides with various functionalities and these aspects have also been reviewed. Solid-phase peptide synthesis is considered as a more established method and has been used to synthesize significant numbers of bioactive peptides [54].

There are several advantages to produce bioactive peptides by chemical synthesis, which include producing peptides of high purity and quantity, desired sequences that are otherwise difficult to obtain from natural sources, and peptides of known activity identified in natural sources but are difficult to ensure their release through hydrolysis. Products such as soy milk, tempeh, and tofu are the examples of soy products consumed as protein sources, whereas soy sauce and miso are examples of soy products used to flavor other food preparations. Major applications of soy protein apart from its use for human consumption are in paper coatings [55] and as animal feed. Currently, however, soybean proteins are used in the preparation of functional peptides such as antioxidative [56–58], antihypertensive [59–61], and anticancer [62] peptides.

Clostripain (EC 3.4.22.8) also known as Arg-C peptidase [63] is a sulfhydryl protease produced by *Clostridium histolyticum* and prefers arginine over lysine residue. Under controlled hydrolysis conditions it will cleave proteins at peptide bonds limited to arginine P1 sites, including trypsin-resistant arginylprolyl bonds [64], and produce peptides containing arginine residues at C1 positions. Many peptides produced from these hydrolyses have also been produced from other food protein sources and reported extensively in scientific literature to have ACE-inhibitory activity. This type of protein has been reported to possess antinutritive action and hydrolysis of this protein has been reported to degrade protease inhibitor in soybean [65]. This degradation has twofold impact, that is, increase the bioavailability of soybean protein and decreasing anticancer and anti-inflammatory effect of this soybean protein that has been previously reported [66]. Various studies have reported the antihypertensive effects of ACE-inhibitory peptides derived from food proteins [67–70].

It is also indicated that small ACE-inhibitory peptides are less susceptible to gastrointestinal enzyme degradation and can be absorbed intact into the blood circulation [71, 72]. Chen *et al.* [73] and Saito *et al.* [74], for example, identified antioxidative peptides from soybean proteins having 5–16 amino acid residues. These peptides consisted of hydrophobic amino acids, valine (V) or leucine (L), at the N-terminal positions, and proline (P), histidine (H) and tyrosine (Y) in any position in the sequences. Many other tripeptides, tetrapeptides and oligopeptides contain sequence of known DPP-IV inhibitors produced by *in silico* hydrolysis of β-conglycinins.

The sequence/s of the active peptide/s remains unknown, the findings do indicate the presence of antioxidative peptides released from the fermentation process of soybean proteins. Hypocholesterolemic activity of peptides derived from fermented soybean product has also been reported. Shon *et al.* [75] reported that fermentation of soy bean in the preparation of soy sauce produces bioactive peptides that can function as antithrombotic agents, although the peptide responsible for this activity remains unknown.

The endopeptidases can be further classified based on their catalytic site into serine proteases, cysteine proteases, metalloproteases, and aspartic proteases [76, 77], while a group of threonine peptidases has also been discovered [78]. The serine proteases are well studied having a maximum activity at alkaline pH and –OH group in the catalytic centre. In the study of high molecular weight fractions (>10 kDa) of hydrolysates, proteins have been found to show higher antioxidant activities when compared to the lower molecular weight fractions [79], although the sequences of active peptides have not been determined. Other researchers, however, have reported various antioxidative peptides derived through hydrolysis of soy proteins [80].

This method has been used successfully to synthesize more than 20 residue peptides. However, this method is time-consuming and the intermediate products need to be isolated and purified before proceeding to the next steps [81]. The solid-phase peptide synthesis is considered as a more established method and has been used to synthesize significant numbers of active peptides. There are various reasons to produce bioactive peptides through chemical synthesis. An extended list of therapeutic proteins produced through genetic engineering using *Chlamydomonas reinhardtii* is available in the literature [82]. The principle of therapeutic protein production in *C. reinhardtii* chloroplast involves the introduction of gold or tungsten particles coated with gene for the foreign protein into the algae cells placed at an interior of a vacuum chamber [83]. Two important ones are antihypertensive and antimicrobial agents. As an antihypertensive agent, the use of bioactive peptides as a food ingredient is advisable [84]. Therefore, the functional properties of these peptides assume importance.

1.7 Soy Protein Isolate-Based Films

Soy protein isolate (SPI) films are clearer, smoother, and more flexible; they also show better gas barrier properties than films made by lipids or polysaccharides [85–87]. SPI is a highly refined or purified form of soy protein – over 90% of its protein content is obtained from defatted SF. SPI has higher protein content and better film-forming ability than other SP products such as SF which has 54% protein, or SPC which has 65–72% protein [88–90]. SPI shows considerable potential for use in food, agriculture, bioscience, and biotechnology industry applications. That said, several problems limit their application in practice, including relatively poor mechanical properties, and poor water resistance and high moisture sensibility compared to petroleum-based plastics. These limitations occur due to the inherent hydrophilicity of the material as well as the strong molecular interactions of natural proteins. Many previous researchers have proposed methods of enhancing SPI-based film performance, including custom processing methods, cross-linking with chemicals like formaldehyde, glyoxal, and glutaraldehyde (GA), phenolic compounds, and epoxy compounds, polymerizing with functional groups, and blending with PGA, PVA, PCL, PLA, PU, chitin, wheat gluten, gelatin, keratin, milk protein, zein, or silks. Zhang *et al.*, suggested a multistep process was used to form SPI sheets by mixing SPI, water, glycerol, and other additives in a high-speed mixer, then applying a twin-screw extrusion to form pellet-shaped material [91].

In a study by Mo *et al.*, the thermal properties of molded SPI films plasticized by four polyol-based plasticizers were analyzed using DMA methods ranging from −30 °C to 200 °C at a rate of 5 °C/min, under a static force of 660 mN and dynamic force of 600 mN at 1 Hz frequency. The SPI/PBAT blends were all stable below 200 °C, and a two-step weight loss was observed between 200 °C and 400 °C in line with the two DTG curve peaks around 220 °C and 310 °C, owing to the processing additives and SPI degradation, respectively. Native SPI exhibits a typical bell shape confirming its spherical morphology; heat-treated SPI samples show decreased magnitude of maximum peaks, suggesting the presence of globular shapes and partially unfolded structures of rigid glycinin and β-conglycinin. Researchers have proposed physical, chemical, and combination modifications in effort to extend the practical application of SPI films and overcome disadvantages such as poor mechanical properties and moisture sensitivity [92, 93]. Blending SPI with other biodegradable polymers is generally accepted to be successful [94, 95]. Blending methods do show favorable reinforcing effects, and even low nanofiller loads produce better SPI-based bionanocomposites than SPI alone [96, 97]. The specialized bio-films formed by SPI modified by Cu NCs exhibit better mechanical properties, and those modified with Zn NCs show better hydrophobic properties, than SPI alone. Further X-ray diffraction testing of SPI films indicated that metal nanoclusters change the conformation of SPI from compact to unfolded. Modified SPI-based films can also be used in biochemical and biomedical research fields. Luo *et al.* conducted a study on a nerve guide conduit from cellulose/SPI hollow tube (CSC) combined with Schwann cells and pyrroloquinoline quinone, and found that the blended film has the ability to repair and reconstruct nerve structure and function owing to the comprehensive contributions from the hollow CSC tube [98]. For the sake of comparison, TiO_2-SPI composite films were also prepared by hot surface casting with TiO_2 nanoparticles as filler [99]. Mechanical and barrier properties (water vapor, and O_2 permeability) of these films were evaluated under different relative humidity (RH) conditions, and in general, TS and E decreased and ε_b increased as RH increased. Fang *et al.* prepared SPI/PLA composites with or without MDI and found that TS of PLA/SPI material without MDI was very low due to the stress concentration on the dispersed SPI. After MDI was added, conversely, the TS of the composites significantly improved [100]. SPI films create semipermeable barriers to water vapor or oxygen between food and its ambient environment, effectively protecting foods from oxidative damage [101–104].

Guerrero *et al.* found that fresh beef patties treated with SPI coating stayed fresh for 14 days as a result of SPI coating delaying the oxidation and deterioration of lipids. The surface color stability of the patties was also maintained during chilled storage [105]. In a similar study by Wang *et al.*, bionanocomposites of SPI/TiO_2 film caused a sizeable decline in water vapor and oxygen permeability (decreased by 72.12% and 57.64%, respectively) compared to the control [106]. The SPI film networks displayed excellent compound binding capacity, especially for hydrophobic molecules, and as such showed attractive potential for use in controlled release systems based on matrix erosion.

1.8 Use of Soy Protein-Based Carriers for Encapsulating Bioactive Ingredients

Currently, proteins extracted from animal-derived products (whey proteins, gelatin, casein) and from vegetables (soy proteins, pea proteins, cereal proteins) have been widely used for encapsulation of bioactive ingredients. Compared with animal-derived proteins, the use of vegetable proteins as encapsulation materials shows the present "green" conception in food industry. Soy proteins are obtained from soy bean contain a fraction (35–40%) of proteins mainly glycinin and conglycin (50–90% of total proteins) [107]. Molecular weight of glycinin fraction (11S globulin) is about 350 kDa while conglycin (7S globulin fraction) is about 70 kDa. Soy proteins have interesting physicochemical and functional properties, particularly including gel-forming, emulsifying, and surfactant properties [108–110]. Hydrolyzed SPI could produce the emulsions with smaller droplets and lower oil retention efficiency (39%) in the corresponding powder due to the insufficient chain length of wall material to produce a stable matrix during spray drying. On the other hand, SPI with *N*-acylation resulted in a significantly higher retention efficiency (>87%) compared to the efficiency obtained with native SPI (80%). SPI microparticles can be made using spray-drying, coacervation, and extrusion [111–113].

Chen and Subirade prepared SPI/zein complex microspheres by extrusion method for delivering riboflavin. The obtained particles (about 15–25 μm) had spherical morphology with homogenous distribution throughout the matrix. Microspheres with SPI/zein ratios of 5:5 and 3:7 displayed near-zero-order release kinetics in simulated gastrointestinal fluids. The results indicated that SPI/zein microspheres showed potential as nutraceutical delivery carriers for the development of functional foods, such as yogurt enriched with vitamins. Rascon *et al.* [114] investigated the performance of SPI for the encapsulation of paprika oleoresin by spray drying. The authors found that carotenoid retention in the microparticles increased as inlet air temperature was increased from 160 °C to 200 °C. Compared to SPI nanoparticles without FA, FA-SPI nanoparticles showed a lower average size, a higher loading efficiency, and a faster release of curcumin in Tween 20-PBS buffer. The results showed that SPI treated by ultrasonics prior to encapsulation could increase their solubility. The optimum conditions for high encapsulation efficiency were followed as: pH 4.0, 1:1 SPI/GA ratio and 10% core material load.

References

1. Vu Huu, T., Kazuo, S., Beta-conglycinin from soybean proteins. Isolation and immunological and physicochemical properties of the monomeric forms. *Biochimica et Biophysica Acta (BBA)*, 490, 370–384, 1977.
2. Wolf, W.J., Soybean proteins. Their functional, chemical and physical properties. *J. Agric. Food. Chem.*, 18, 969–976, 1970.
3. Ali, F., Ippersiel, D., Lamarche, F., Mondor, M., Characterization of low-phytate soy protein isolates produced by membrane technologies. *Innov. Food Sci. Emerg. Technol.*, 11, 162–168, 2010.
4. Malhotra, A., Coupland, J.N., The effect of surfactants on the solubility, zeta potential, and viscosity of soy protein isolates. *Food Hydrocoll.*, 18, 101–108, 2004.
5. Deak, N.A., Murphy, P.A., Johnson, L.A., Fractionating soybean storage proteins using Ca^{2+} and $NaHSO_3$. *J. Food Sci.*, 71, C413–C424, 2006.
6. Kumar, R., Choudhary, V., Mishra, S., Varma, I.K., Mattiason, B., Adhesives and plastics based on soy protein products. *Ind. Crops Prod.*, 16, 155–172, 2002.
7. Van de Weert, M., Stella, L., Fluorescence quenching and ligand binding: a critical discussion of a popular methodology. *J. Mol. Struct.*, 998, 144–150, 2011.
8. Qu, P., Huang, H., Wu, G., Sun, E., Chang, Z., The effect of hydrolyzed soy protein isolate on the structure and biodegradability of urea-formaldehyde adhesives. *J. Adhes. Sci. Technol.*, 2015.
9. Khatib, K.A., Herald, T.J., Aramouni, F.M., MacRitchie, F., Schapaugh, W.T., Characterization and Functional Properties of Soy β-Conglycinin and Glycinin of Selected Genotypes. *J. Food Sci.*, 67(8), 2923–2929, 2002.
10. Zhang, X., Burgar, I., Do, M.D., Lourbakos, E., Intermolecular interactions and phase structures of plasticized wheat proteins materials. *Biomacromolecules*, 6(3), 1661–1671, 2005.
11. Ly, Y.T.-P., Johnson, L.A., Jane, J., Soy protein as biopolymer. In *Biopolymers from renewable Resources*, D.L. Kaplan, Ed., pp. 144–176, Springer Berlin Heidelberg: Berlin, Heidelberg, 1998.
12. Chen, P., Zhang, L., Cao, F., Effects of moisture on glass transition and microstructure of glycerol-plasticized soy protein. *Macromol. Biosci.*, 5, 2005.
13. González, A., Alvarez Igarzabal, C.I., Nanocrystal-reinforced soy protein films and their application as active packaging. *Food Hydrocoll.*,43, 777–784, 2015.
14. Kumar, R., Choudhary, V., Mishra, S., Varma, I.K., Mattiason, B., Adhesives and plastics based on soy protein products. *Ind. Crops Prod.*,16(3), 155–172, 2002.
15. Guo, G., Zhang, C., Du, Z., Zou, W., Xiang, A., Li, H., Processing and properties of phthalic anhydride modified soy protein/glycerol plasticized soy protein composite films. *J. Appl. Polym. Sci.*, 132(28), 2015.

16. Sasmal, A., Sahoo, D., Nanda, R., Nayak, P., Nayak, P.L., Mishra, J.K., Chang, Y.-W., Yoon, J.-Y., Biodegradable nanocomposites from maleated polycaprolactone/soy protein isolate blend with organoclay: Preparation, characterization, and properties. *Polym. Composites*, 30(6), 708–714, 2009.
17. Fujin A., Hua Z., Ming W., Jin, H., Soy protein plastics reinforced and toughened by SiO_2 nanoparticles. *J. Appl. Polym. Sci.*, 105(3), 8, 2007.
18. Zheng, H. F. Thermoplastic soy protein nanocomposites reinforced by carbon nanotubes. *Macromol. Mater. Eng.*, 292, 2007.
19. Sasmal,., Biodegradable nanocomposites from maleated polycaprolactone/soy protein isolate blend with organoclay: preparation, characterization, and properties. *Polym. Composites*, 30(6), 7, 2009.
20. Yang, S., Leong, K.F., Du, Z., Chua, C.K., The design of scaffolds for use in tissue engineering. Part I. Traditional factors. *Tissue Eng.*, 7, 679–689, 2001.
21. Vaz, C.M., Van Doeveren, P.F., Reis, R.L., Cunha, A.M., Soy matrix drug delivery systems obtained by melt-processing techniques. *Biomacromolecules*, 4, 1520–1529, 2003a.
22. Angellier, H., Molina-Boisseau, S., Dole, P., Dufresne, A., Thermoplastic starch-waxy maize starch nanocrystals nanocomposites. *Biomacromolecules*, 7, 531–539, 2006.
23. Chen, P., Zhang, L., Interaction and properties of highly exfoliated soy protein/montmorillonite nanocomposites. *Biomacromolecules*, 7, 1700–1706, 2006.
24. Regalado, C., Perez-Perez, C., Lara-Cortes, E., Garcia-Almendarez, B., Guevara-González, R.G., Torres-Pacheco I. "Whey protein based edible food packaging films and coatings." *Adv. Agric. Food Biotechnol.*, 237–261, 2006.
25. Meikle, S.T., Standen, G., Salvage, J., De Santis, R., Nicolais, L., Ambrosio, L., Santin, M., Synthesis and characterization of soybean-based hydrogels with an intrinsic activity on cell differentiation. *Tissue Eng. Part A18*, 1932–1939, 2012.
26. Wang, Y., Cao, X., Zhang, L., Effects of cellulose whiskers on properties of soy protein thermoplastics. *Macromol. Biosci.*, 6, 524–531, 2006.
27. Gennadios, A., Brandenburg, A.H., Weller, C.L., Testin, R.F., Effect of pH on properties of wheat gluten and soy protein isolate films. *J. Agric. Food Chem.*, 41, 1835–1839, 1993.
28. Li, K., Peshkova, S., Geng, X., Investigation of soy protein-Kymene® adhesive systems for wood composites. *J. Am. Oil Chem. Soc.*, 81, 487–491, 2004.
29. Thames, S.F., Cook, R.C., Mendon, S.K., U.S. Patent No. 6,790,271. Washington, DC: U.S. Patent and Trademark Office, 2004.
30. Rhim, J.W., Gennadios, A., Handa, A., Weller, C.L., Hanna, M.A., Solubility, tensile, and color properties of modified soy protein isolate films. *J. Agric. Food Chem.*, 48, 4937–4941, 2000.
31. Luo, L.H., Wang, X.M., Zhang, Y.F., Liu, Y.M., Chang, P.R., Wang, Y., Chen, Y., Physical properties and biocompatibility of cellulose/soy protein isolate membranes coagulated from acetic aqueous solution. *J. Biomater. Sci., Polym. Ed.*, 19, 479–496, 2008.

32. Vucenik, I., Shamsuddin, A.M., Cancer inhibition by inositol hexaphosphate (IP6) and inositol: from laboratory to clinic. *J. Nutr.*, 133, 3778S–3784S, 2003.
33. Yoon, J.H., Thompson, L.U., Jenkins, D.J.A., The effect of phytic acid on *in vitro* rate of starch digestibility and blood glucose response. *Am. J. Clin. Nutr.*, 38, 835–842, 1983.
34. Sudheer, K.M., Sridhar, R.B., Kiran, B.S., Bhilegaonkar, P.M., Shirwaikar, A., Unnikrishnan, M.K., Anti-inflammatory and anti-ulcer activities of phytic acid in rats. *Ind. J. Exper. Biol.*, 42, 179–185, 2004.
35. Lindsay, S.H., Claywell, L.G., Considering soy: its estrogenic effects may protect women. *AWHONN Lifelines*, 2, 41–44, 1998.
36. Ramnath, V., Sekar, S., Sankar, S., Sastry, T.P., Mandal, A.B., *In vivo* evaluation of composite wound dressing material containing soya protein and sago starch. *Int. J. Pharm. Pharm. Sci.*, 4, 1–6, 2012.
37. Chien, K.B., Aguado, B.A., Bryce, P.J., Shah, R.N., *In vivo* acute and humoral response to three-dimensional porous soy protein scaffolds. *Acta Biomater.*, 9, 8983–90, 2013.
38. Chien, K.B., Shah, R.N., Novel soy protein scaffolds for tissue regeneration: material characterization and interaction with human mesenchymal stem cells. *Acta Biomater.*, 8, 694–703, 2012.
39. Xu, H., Cai, S., Sellersc, A., Yang, Y., Intrinsically water-stable electrospun three dimensional ultrafine fibrous soy protein scaffolds for soft tissue engineering using adipose derived mesenchymal stem cells. *RSC Adv.*, 4, 15451–15457, 2014.
40. Silva, G.A., Vaz, C.M., Coutinho, O.P., Cunha, A.M., Reis, R.L., *In vitro* degradation and cytocompatibility evaluation of novel soy and sodium caseinate-based membrane biomaterials. *J. Mater. Sci. Mater. Med.*, 14, 1055–1066, 2003.
41. Bhatia, S.K., Bio-based materials steps into the operating room. www.aiche.org/cep, 49–53, 2012. Accessed 18th February 2015.
42. Vaz, C.M., van Doeveren, P.F.N.M., Reis, R.L., Cunha, A.M., Soy matrix drug delivery systems obtained by melt-processing techniques. *Biomacromolecules*, 4, 1520–1529, 2003.
43. Siva Mohau Reddy, G., Jayaramudu, J., Sadiku, E.R., Shina Ray, S., Varaprasad, K., Aderibigbe, B.A., Application of cross-linked soy protein isolate with resorcinol films for drug release studies of naturally occurring bioactive agent with antiproliferative activity, *J. Drug Del. Sci. Tech.*, 24, 86–91, 2014.
44. Chien, K.B., Chung, E.J., Shah, R.N., Investigation of soy protein hydrogels for biomedical applications: materials characterization, drug release, and biocompatibility. *J. Biomater. Appl.*, 28, 1085–1096, 2014.
45. Kinsella, J., *J. American. Oil Chem. Soc.*, 56, 242–258, 1979.
46. Peng, I.C., Quass, D.W., Dayton, W.R., Allen, C.E., *Cereal Chem.*, 61, 480–490, 1984.
47. Lubasova, D., Netravali, A., Parker, J., Ingel, B., *J. Nanosci. Nanotechnol.*, 14, 4891–4898, 2014.
48. Lubasova, D., Mullerova, J., Netravali, A.N., *J. Appl. Polym. Sci.*, 132, 41852, 2015.

49. Salas, C., Ago, M., Lucia, L.A., Rojas, O.J., *React. Funct. Polym.*, 85, 221–227, 2014.
50. Phiriyawirut, M., Rodchanacheewa, N., Nensiri, N., Supaphol, P., *Adv. Mater. Res.*, 2008.
51. Cho, D., Netravali, A.N., Joo, Y.L., *Polym. Degrad. Stab.*, 97, 747–754, 2012.
52. Xu, X., Jiang, L., Zhou, Z., Wu, X., Wang, Y., *ACS Appl. Mater. Interfaces*, 4, 4331–4337, 2012.
53. Thirugnanaselvam, M., Gobi, N., Arun Karthick, S., *Fibers Polym.*, 14, 965-969, 2013.
54. Guzmán, F., Barberis, S., Illanes, A., *Electron. J. Biotechnol.*, 10, 279–314, 2007.
55. Johnson, L.A., Myers, D.J., Burden, D.J., *Inform*, 3, 282–290, 1992.
56. Singh, B.P., Vij, S., Hati, S., *Peptides*, 54, 171–179, 2014.
57. Wang, Y.C., Yu, R.C., Chou, C.C., *Food Microbiol.*, 23, 128–135, 2006.
58. Zhao, J., Xiong, Y.L., *LWT - Food Sci. Technol.*, 61, 322–329, 2015.
59. De Mejia, E., De Lumen, B.O., *Sex., Reprod. Menopause*, 4, 91–95, 2006.
60. Puchalska, P., Concepción García, M., Luisa Marina, M., *Food Chem.*, 157, 62–69, 2014.
61. Tsai, J.S., Lin, Y.S., Pan, B.S., Chen, T.J., *Process Biochem.*, 41, 1282–1288, 2006.
62. Kim, S.K., Kim, H.H., Kim, J.Y., Kang, Y.I., Woo, H.J., Lee, S.E., *Biofactors*, 12, 151–155, 2000.
63. Banèeres, J.-L., Essalouh, L., Jariel-Encontre, I., Mesnier, D., Garrod, S., Parello, J., *J. Mol. Biol.*, 243, 48–59, 1994.
64. Mitchell, W.M., In: *Methods in enzymology*, Vol. 47, Academic Press, pp. 165–170, 1977.
65. Hernández-Ledesma, B., Hsieh, C.C., de Lumen, B.O., *Food Chem.*, 115, 574–580, 2009.
66. Ribeiro, J.K., Cunha, D.D., Fook, J.M., Sales, M.P., *Eur. J. Pharmacol.*, 644, 238–244, 2010.
67. Nakamura, K., Naramoto, K., Koyama, M., *J. Funct. Foods*, 5, 406–415, 2013.
68. Ngo, D.H., Kang, K.H., Jung, W.K., Byun, H.G., Kim, S.K., *J. Funct. Foods*, 10, 243–251, 2014.
69. Balti, R., Bougatef, A., Sila, A., Guillochon, D., Dhulster, P., Nedjar-Arroume, N., *Food Chem.*, 170, 519–525, 2015.
70. Jao, C.L., Huang, S.L., Hsu, K.C., *Biomedicine*, 2, 130–136, 2012.
71. Wu, J., Aluko, R.E., Nakai, S., *J. Agric. Food Chem.*, 54, 732–738, 2006.
72. Foltz, M., Meynen, E., Bianco, V., van Platerink, C., *J. Nutr.*, 137, 953, 2007.
73. Chen, H.M., Muramoto, K., Yamauchi, F., *J. Agric. Food Chem.*, 43, 574–578, 1995.
74. Saito, K., Jin, D.H., Ogawa, T., Muramoto, K., Hatakeyama, E., Yasuhara, T., Nokihara, K., *J. Agric. Food Chem.*, 51, 3668–3674, 2003.
75. Shon, D.H., Lee, K.E., Ahn, C.W., Nam, H.S., Lee, H.J., Shin, J.K., *Korean J. Food Sci. Technol.*, 28, 684–688, 1996.

76. Whitaker, J.R., In: *Handbook of food enzymology*, J.R. Whitaker, A.G.J. Voragen, D.W.S. Wong, Eds., Marcel Dekker Inc., New York, 2003a.
77. Whitaker, J.R., In *Handbook of food enzymology*, J.R. Whitaker, A.G.J. Voragen, D.W.S. Wong, Eds., Marcel Dekker Inc., New York, 2003b.
78. Barret, A.J., In: *Proteolytic enzymes*, R. Beynon, J.S. Bond, Eds., Oxford University Press, New York, 2nd edn., 2001.
79. Moure, A., Domínguez, H., Parajó, J.C., *Process Biochem.*, 41, 447–456, 2006.
80. Jiménez-Escrig, A., Alaiz, M., Vioque, J., Rupérez, P., *Eur. Food Res. Technol.*, 1–9, 2009.
81. Carpino, L.A., Ghassemi, S., Ionescu, D., Ismail, M., Sadat-Aalaee, D., Truran, G.A., Mansour, E.M.E., Siwruk, G.A., Eynon, J.S., Morgan, B., *Org. Process Res. Develop.*, 7, 28–37, 2003.
82. Almaraz-Delgado, A.L., Flores-Uribe, J., Pérez-España, V.H., Salgado-Manjarrez, E., Badillo-Corona, J.A., *AMB Exp.*, 4, 1–9, 2014.
83. Glick, B.R., *Molecular biotechnology : principles and applications of recombinant DNA*, 4th edn., ASM Press, Washington, DC, 2010.
84. Matsui, T., Matsufuji, H., Seki, E., Osajima, K., Nakashima, M., Osajima, Y., *Biosci. Biotechnol. Biochem.*, 57, 922–925, 1993.
85. Kunte, L.A., Gennadios, A., Cuppett, S.L., Hanna, M.A., Weller, C.L., Cast films from soy protein isolates and fractions 1. *Cereal Chem.*,74, 115–118, 1997.
86. Cho, S.Y., Rhee, C., Mechanical properties and water vapor permeability of edible films made from fractionated soy proteins with ultrafiltration. *LWT-Food Sci. Technol.*, 37, 833–839, 2004.
87. Gennadios, A., Brandenburg, A.H., Weller, C.L., Testin, R.F., Effect of pH on properties of wheat gluten and soy protein isolate films. *J. Agric. Food Chem.*, 41, 1835–1839, 1993.
88. Xu, F.J., Zhang, W., Zhang, S.F., Li, L., Li, J.Z., Zhang, Y., Preparation and characterization of poly(vinyl alcohol) and 1,2,3-propanetriol diglycidyl ether incorporated soy protein isolate-based films. Journal of Applied Polymer Science,132, 42578–80, 2015.
89. Xu, F.J., Dong, Y.M., Zhang, W., Zhang, S.F., Li, L., Li, J.Z., Preparation of cross-linked soy protein isolate-based environmentally-friendly films enhanced by PTGE and PAM. *Ind. Crops Prod.*, 67, 373–380, 2015.
90. Hammann, F., Schmid, M., Determination and quantification of molecular interactions in protein films: a review. *Materials*, 7, 7975–7996, 2014.
91. Zhang, J., Mungara, P., Jane, J.-l., Mechanical and thermal properties of extruded soy protein sheets. *Polymer*, 42, 2569–2578, 2001.
92. Tummala, P., Liu, W., Drzal, L.T., Mohanty, A.K., Misra, M., Influence of plasticizers on thermal and mechanical properties and morphology of soy-based bioplastics. *Ind. Eng. Chem. Res.*, 45, 7491–7496, 2006.
93. Liu, W., Misra, M., Askeland, P., Drzal, L.T., Mohanty, A.K., 'Green' composites from soy based plastic and pineapple leaf fiber: fabrication and properties evaluation. *Polymer*,46, 2710–2721, 2005.

94. Zhang, J.W., Jiang, L., Zhu, L.Y., Jane, J.l., Mungara, P., Morphology and properties of soy protein and polylactide blends. *Biomacromolecules*, 7, 1551–1561, 2006.
95. Huang, J., Zhang, L.N., Chen, F.G., Effects of lignin as a filler on properties of soy protein plastics. 1. Lignosulfonate. *J. Appl. Polym. Sci.*, 88, 3284–3290, 2003.
96. Wei, M., Fan, L.H., Huang, J., Chen, Y., Role of star-like hydroxylpropyl lignin in soy-protein plastics. *Macromol. Mater. Eng.*, 291, 524–530, 2006.
97. Wang, Y.X., Cao, X.D., Zhang, L.N., Effects of cellulose whiskers on properties of soy protein thermoplastics. *Macromol. Biosci.*, 6, 524–531, 2006.
98. Luo, L., Gan, L., Liu, Y., Tian, W., Tong, Z., Wang, X., *et al.*, Construction of nerve guide conduits from cellulose/soy protein composite membranes combined with Schwann cells and pyrroloquinoline quinone for the repair of peripheral nerve defect. *Biochem. Biophys. Res. Commun.*, 457, 507–513, 2015.
99. Jensen, A., Lim, L.T., Barbut, S., Marcone, M., Development and characterization of soy protein films incorporated with cellulose fibers using a hot surface casting technique. *LWT - Food Sci. Technol.*,60, 162–170, 2015.
100. Fang, K., Wang, B., Sheng, K., Sun, X.S., Properties and morphology of poly(lactic acid)/soy protein isolate blends. *J. Appl. Polym. Sci.*, 114, 754–759, 2009.
101. McHugh, T.H., Krochta, J.M., Sorbitol-vs glycerol-plasticized whey protein edible films: integrated oxygen permeability and tensile property evaluation. *J. Agric. Food Chem.*, 42, 841–845, 1994.
102. Valencia-Chamorro, S.A., Palou, L., Del Río, M.A., Pérez-Gago, M.B., Antimicrobial edible films and coatings for fresh and minimally processed fruits and vegetables: a review. *Crit. Rev. Food Sci. Nutr.*, 51, 872–900, 2011.
103. Falguera, V., Quintero, J.P., Jiménez, A., Muñoz, J.A., Ibarz, A., Edible films and coatings: structures, active functions and trends in their use. *Trends in Food Sci. Technol.*, 22, 292–303, 2011.
104. González A, Igarzabal CIA. Nanocrystal-reinforced soy protein films and their application as active packaging. Food Hydrocoll. 2015;43:777-84.
105. Guerrero, P., O'Sullivan, M.G., Kerry, J.P., de la Caba, K., Application of soy protein coatings and their effect on the quality and shelf-life stability of beef patties. *RSC Adv.*, 5, 8182–8189, 2015.
106. Wang, Z., Zhang, N., Wang, H.-y., Sui, S.-y., Sun, X.-x., Ma, Z.-s., The effects of ultrasonic/microwave assisted treatment on the properties of soy protein isolate/titanium dioxide films. LWT - Food Science and Technology, 57, 548–555, 2014.
107. Rui-Henestrosa, V.P., Sanchez, C.C., Escobar, M.M.Y., Jimenez, J.J.P., Rodriguez, F.M., Patino, J.M.R., Interfacial and foaming characteristics of soy globulins as a function of pH and ionic strength. *Colloid. Surf. B Biointerf.*, 309, 202–215, 2007.

108. Nesterenko, A., Alric, I., Silvestre, F., Durrieu, V., Vegetable proteins in microencapsulation: a review of recent interventions and their effectiveness. *Ind. Crops Prod.*, 42, 469–479, 2013.
109. Wan, Z.L., Guo, J., Yang, X.Q., Plant protein-based delivery systems for bioactive ingredients in foods. *Food Funct.*, 6, 2876–2889, 2015.
110. Gu, X., Campbell, L.J., Euston, S.R., Effects of different oils on the properties of soy protein isolate emulsions and gels. *Food Res. Int.*, 42, 925–932, 2009.
111. Lazko, J., Popineau, Y., Renard, D., Legrand, J., Microcapsules based on glycinin-sodium dodecyl sulfate complex coacervation. *J. Microencapsul.*, 21, 59–70, 2004.
112. Chen, L., Subirade, M., Elaboration and characterization of soy/zein protein microspheres for controlled nutraceutical delivery. *Biomacromolecules,* 10, 3327–3334, 2009.
113. Chen, L., Hébrard, G., Beyssac, E., Denis, S., Subirade, M., *In vitro* study of the release properties of soy–zein protein microspheres with a dynamic artificial digestive system. *J. Agric. Food Chem.*, 58, 9861–9867, 2010.
114. Rascon, M.P., Beristain, C.I., Garcie, H.S., Salgado, M.A., Carotenoid retention and storage stability of spray-dried paprika oleoresin using gum Arabic and soy protein isolate as wall materials. *Food Sci. Technol.,* 44, 549–557, 2011.

2

Soy Protein: Introduction, Structure and Properties Relationship

Visakh P. M

Department of Ecology and Basic Safety, Tomsk Polytechnic University, Tomsk, Russia

Abstract

This chapter gives an information on the general aspects of soy protein introduction, structure, information, and properties. Chemical modification and characterization of soy proteins are also included in this chapter. This chapter also describes the applications of soy protein-based nanocomposites and blends.

***Keywords*:** Extraction of soy proteins, 7S soy proteins, inositol hexakisphosphate, emulsifying activities of soy protien, 11S globulin

2.1 Introduction

Soy proteins are one of the most abundant and most widely utilized plant proteins on this planet. They include a number of globulins which represent approximately 30–40% of the total weight of soybean. This chapter will describe the source and production of soy proteins, then after discussing about structural and functional properties.

2.2 Structure of Soy Proteins

Soy proteins represent a group of globulins which are conventionally classified into four major components, that is, 2S, 7S, 11S, and 15S globulins.

Corresponding author: visagam143@gmail.com

Visakh P. M and Olga Nazarenko (eds.) Soy Protein-Based Blends, Composites and Nanocomposites, (23–38) 2017 © Scrivener Publishing LLC

The 11S and 7S globulins represent approximately 60% of the storage protein in soybeans. They are the major components of soy protein isolate (SPI), the most common soy protein product [1]. The 11S globulin (glycinin) is a heterogenic protein complex with a MW of 320,000 Da. Six subunits are present in the glycinin molecule, each of which contains one acidic and one basic peptide linked via a disulphide bond. The 7S globulin contains β-conglycinin as the major component which accounts for more than a half of this fraction. The β-conglycinin shows an average MW of 180,000 Da [2] and pI of 4.8. β-conglycinin is relatively flexible, as evidenced by its high contents of α-helix, and random coils [3].

It comprises six major isomers (designated as B1- to B6-conglycinin), each of which are composed of three major subunits (α, α', and β, with the MW of 57, 57, and 42 kDa, respectively) and two minor ones (γ and δ) [4]. The major carbohydrates present in β-conglycinin are mannose (2.9~4.1% by weight) and glucosamine (1.0–1.2% by weight), whose contents vary among different isomers. The 2S globulins including trypsin inhibitor and cytochrome c make up 8~20% of the soybean storage protein [5, 6]. Aggregation form of β-conglycinin is strongly dependent on the ionic strength and pH. Glycinin has been showed with a similar tendency of dimerization at low ionic strength, although the proportion of dimers is not as high as for β-conglycinin. Glycinin tends to dissociate into 2S, 3S, or 7S forms under different ionic strengths and pHs [7]. Therefore, environmental factors must be taken into account when preparing soy protein-based formulations and materials. These proteins are largely removed in the preparation of SPI, which involves the precipitation at acidic pHs.

2.3 Source of Soy Proteins

Whole beans contain approximately 40.4% crude proteins. The soybean seeds can be processed into different forms such as soy meal, soy flour, and soy flakes, all of which could serve as a starting material for soy protein extraction. As can be seen, the whole soybean seed is cleaned, cracked, dehulled, and flaked to produce soy powder. The insoluble fraction containing mainly proteins and carbohydrates is evaporated to yield defatted soy meal. The typical particle size of the fine soy flour is less than 100-mesh or 150 μm. In general, smaller particles possess higher specific surface area, which usually lead to a higher recovery of soy protein regardless of the type of preparative method. Functional properties such as solubility and emulsifying ability could also vary greatly among different genotypes of soybeans [8].

2.3.1 Extraction of Soy Proteins

There exist three major types of soy protein-rich products, namely SPI (purity>90%), SPC (soy protein concentrate, purity>70%), and fractionated 11S/7S globulins (protein content>90%, fraction purity>85%). It should be pointed out that the structure of soy proteins is dynamic and subject to change at different conditions (e.g., pH, ionic strength, reducing agents, temperature). In addition, heat and reducing agents also have a significant impact on the structure of soy proteins, leading to extensive aggregation and dissociation, respectively.

2.3.2 Production of SPI

SPI is one of the most enriched forms of soy protein products, exhibiting a typical protein content of 90% or above and protein dispersion index (PDI) of 80–90% [9]. The majority of the protein is precipitated and recovered by a second centrifugation. The typical protein of yield (weight ratio between the product and the raw material) is around 30%, though a yield of as high as 44% has been reported [10]. Ali and colleagues [11] proposed the preparation of SPI using a membrane-based method. The product demonstrated similar solubility to that of the isoelectric-precipitated counterpart, and it contained much less phytic acid, the major source of phosphorous and an antinutritional factor.

2.3.3 Preparation of SPC

The preparation processes usually involve heat or organic solvent treatment, which denature the protein either completely or partly. Three methods differing in the extraction media are conventionally adopted to produce SPC. The soluble fraction containing mostly carbohydrates is removed, and the insoluble part is recovered as the product. Soaking with diluted acid leads to a milder denaturation compared to the other treatments, leading to better solubility of the product. Shallo *et al.* reported that the protein recovery achieved by ultrafiltration was up to 26% higher than that for conventional preparation methods [12].

2.3.4 Preparation of 11S and 7S Soy Proteins

One of the most widely adopted procedures for obtaining pure 11S and 7S soy globulins is based on Nagano's method [13]. Soy proteins are firstly extracted under basic conditions and separated from the fibre through centrifugation. The 11S fraction is precipitated out, and an intermediate fraction is salted out from the resulting supernatant by adding 0.25 M of NaCl and adjusting the pH to 5.0. Both 11S and 7S fractions are neutralized, washed with water, and freeze-dried. 7S protein can then be precipitated directly by acidifying the remaining supernatant [14]. Teng *et al.* investigated the effect of divalent cations on the fractionation process.

2.4 Properties of Soy Proteins

These characteristics are in turn strongly dependent on the interplay between the chemical structure of the protein and environmental factors. The effects of pH, temperature, ionic strength depend on the soy proteins properties. Considering the fact that SPI is the purest form of soy protein product by far, the following sections will be focused on the isolate.

2.4.1 Surface Properties

The surface charge, hydrophobicity, and charge density are critical for stability and reactivity of soy proteins with respect to the synthesis of soy protein-based materials (e.g., nanocomposites). These properties are mainly determined by the structural properties of the soy proteins, as well as environmental factors such as pH and ionic strength. PI of soy proteins is also affected by the coexisting compounds, especially the phosphorous-abundant chemical named phytic acid or inositol hexakisphosphate. Removal of phytic acid by methods such as phytase treatment could greatly reduce the amount of negative charge on soy proteins, shifting its pH to more basic values (e.g., pH 5.0 or above). The preparation method for soy proteins has a significant effect on the phytic acid content, therefore altering the surface charge of the protein products. The surface hydrophobicity of a protein indicates the extent to which its hydrophobic amino acid residues are exposed to the solvent. Value is relatively low compared to other proteins such as β-lactoglobulin, casein, and bovine serum albumin [15], indicating that the majority of the hydrophobic amino acids is buried in the core region of soy proteins. The surface hydrophobicity is higher as the pH is further away from the pI, possibly because the increased electrostatic repulsion results in a more expanded protein conformation and facilitates the exposure of more hydrophobic amino acids [16].

2.4.2 Solubility

As a mixture of 7S and 11S proteins, SPI exhibits a similar solubility profile, displaying minimal (close to zero) percentage of soluble fractions at pH 4.5 and maximal (above 90%) values at pH 2~3 or pH 8~10. Renkema *et al.* studied the influence of ionic strength on the solubility of heated or unheated SPI [17]. Furthermore, the type of ions present in the protein solution also influences the protein solubility. Salt bridging effects between negatively charged soy proteins by the divalent cations [18].

2.4.3 Emulsifying Activities

Emulsion is a homogeneous mixture of immiscible liquids (i.e., water and oil). Emulsification is a vital feature of a polymer in the situation where two or more immiscible reactants need to be blended for efficient reaction. The emulsifying ability is defined as the maximum oil quantity which can be emulsified by a fixed amount of a protein, and the emulsification stability has been defined operationally by the velocity of phase separation into water and oil during storage of emulsion. The emulsifying properties of a protein are dependent on many factors including protein concentration, solubility, and surface hydrophobicity. There is a pronounced difference in the emulsifying and foaming abilities between 11S and 7S soy proteins. Another relevant study suggested that the acidic peptide of the 11S adsorbed to the air-water interface at a higher rate compared to the integrate 11S globulin [19]. The emulsifying activity and emulsification stability of the 7S protein are more than 200% higher than that of 11S [20]. Given the fact that both 11S and 7S soy proteins show equally low surface hydrophobicity. Salt concentration affects the solubility and surface hydrophobicity of soy proteins as described above, thus exerting certain effect on its emulsifying abilities.

2.4.4 Viscoelastic Properties

The aqueous dispersion of soy proteins exhibits the lowest viscosity near the pI (pH = 4.5) [21]. This could be explained by the compact conformation of soy proteins because of the lack of electrostatic repulsion, which reduces the opportunity of interacting with water molecules via hydrogen bonding. Gelation is a complex process that involves the partial unfolding of the protein molecules, followed by the formation of network structure driven by different types of intermolecular associative effects, such as electrostatic attraction (between oppositely charged ions), hydrophobic interaction, van der Vaal force, disulphide bonds, and other covalent bonds produced by enzymes if applicable. Compared to the 7S protein, the 11S fraction demonstrates a shorter initiation time for gelation, as well as significant higher gel elasticity. The existence and chemical reaction involving the disulphide bond in the 11S protein, together with the predominance of rigid β-sheet structure may account for such difference. In the meantime, the tangent of phase angle ($\tan \delta = G''/G'$) of the 11S gel is several orders in magnitudes lower than for the 7S, which suggests that elasticity is much more predominant in the 11S gel [22]. Environmental factors such as temperature, pH, ionic strength, and type of salts have a marked effect on the viscoelastic property of soy proteins.

2.4.5 Thermal Properties

As thermal treatments induce conformational change on the soy proteins, they can further alter the supramolecular structure, producing protein aggregates of different sizes depending on the heating condition. At temperatures higher than 90 °C, the 11S protein forms insoluble aggregates that consist exclusively of the basic peptide. Hydrophobic interaction is the chief driving force for such aggregation. Interestingly, 11S does not form aggregates when incubated together with the 7S fraction.

2.4.6 Thermoplastic Properties

Soy proteins exhibit a relatively high T_g (70 °C), which makes them difficult for moulding and manufacturing. At low moisture contents, water can act as an antiplasticizer by enhancing the structural orderliness of the protein. Other functions of the plasticizer include (1) conferring flexibility by reducing the cohesive intermolecular forces and increasing molecular spacing and (2) improving the adhesive properties [23, 24]. The insertion of glycerol into the peptide chains in the soy proteins and the latter one promote its interaction with the protein via extensive hydrogen bonding.

2.4.7 Film-Forming Abilities

A plasticizer such as glycerol is usually introduced to facilitate the formation of a smooth film. A more recent study suggested that the replacement of glycerol by stearic acid resulted in a marked decrease in the moisture content of soy protein films. The films formed by soy proteins exhibit similar TS and water vapor permeability with those prepared with wheat gluten, a water-insoluble protein, while the *E*% of soy protein film is inferior to the latter [25].

2.5 Chemical Modification of Soy Proteins

Chemical modification is an effective approach to enhance the functional properties of a molecule or to impart it with new properties. Phosphorylation is the chemical attachment of phosphate groups (PO_4^{3-}). Such process confers the protein with a large amount of negative charges, making it more anionic under neutral and basic conditions. The product exhibited much improved functional properties such as aqueous solubility, water-holding capacity, emulsifying ability, and whippability. Dephosphorylation removes the phosphate-rich components bound to the soy proteins. As introduced in previous sections, phytic acid is the major source of phosphorous in soy protein products. This property is favorable for the application of soy proteins in acidic environments. Several methods have been employed to eliminate phytic acid in the soy proteins, including alkali treatment followed by centrifugation [26], ionic exchange process [27], and phytase treatment [28]. The reaction between the positively charged amine groups on the soy proteins and an external carboxylic acid is comparable to phosphorylation. Phosphorylation process which provides additional electric charges to the protein, the amidation process deprives the original charges of the protein, and it adds a relatively hydrophobic hydrocarbon chain originating from the carboxylic acid.

Deamidation is the cleavage of an amide bond into a carboxylic acid and an amine. Soy proteins possess glutamine and asparagine that contain an amide residue. Due to the stability of amides, an enzyme such as *Bacillus circulans* peptidoglutaminase is required for effective deamidation. Phosphorylation, deamidation provides soy proteins with additional negative charges and improves its solubility at acidic pH. Other benefits include better emulsifying capacity as reported previously. Enzymes are useful tools for modifying the structure then enhancing the properties of proteins. Proteolytic enzymes such as trypsin, rennin, and alcalase break down the protein into smaller peptide segments, exposing more hydrophobic and other functional groups to the solvent. The changes in such properties depended upon the enzymes used, and duration of proteolytic treatment. This enzyme aids the formation of an isopeptide bond between a free amine group (e.g., the ε-amine group on lysine).

2.5.1 Chemical Modification Through Oxidation or Reduction

2.5.1.1 Glycosylation

The heat stability is generally improved upon glycosylation possibly due to the shielding effect of the sugars, although the extent of improvement is strongly dependent on the repeating unit and MW of the polysaccharides. The functional properties of glycerol-plasticized soy protein films can be altered by the modification with sodium dodecyl sulphate (SDS). SDS does not cause the formation or breakdown of any covalent bonds; instead, it disrupts the hydrophobic interaction within the protein molecules with its own hydrophobic tails. Chemical reagents such as urea and guanidine hydrochloride denature the protein by substantially disrupting the hydrogen bonding, thus unfolding the protein molecules effectively. The exposure of hydrophobic sites upon protein unfolding results in superior water resistance [29].

2.6 Characterization of Soy Proteins

The application of advanced analytical techniques provides possibility for exploring the change in these properties during the extraction, modification, and processing. Infrared (IR) spectroscopy is a useful tool for characterizing the chemical bonds in a molecule. When an infrared wave is shed on a molecule, the portions of certain wavelengths are absorbed by its chemical bonds. The peaks appearing in the absorbance spectra provide adequate information on the type and rough content of bonds the analyte possesses. Fourier-transform infrared (FT-IR) spectroscopy enables the data collection over a wide range of wavelengths at a same time, and it also provides much higher signal-to-noise ratio [30]. The amide I band at around 1650 cm^{-1} is of particular interest for the investigation of protein structure. Based on many previous literatures, these major bands in this amide I region can be assigned as follows: 1691 cm^{-1}, β-turn and bend; 1674 cm^{-1}, extended strands (β-strand); 1667 cm^{-1} and 1660 cm^{-1}, β-turns; 1652 cm^{-1}, α-helix; 1644 cm^{-1}, unordered structures (random coils); 1635 cm^{-1}, antiparallel β-sheet structure [31]. The characteristic peaks of two amino acid residues (cysteine and protonated glutamic or aspartic acid) are free from overlapping with the hydroxyl and amide I~III bands.

This protein is rich in lysine and hydrophobic amino acids (e.g., leucine and phenylalanine), but it lacks methionine, glutamic acid, and aspartic acid. FT-IR revealed its abundance in rigid structures such as β-sheets and β-turns. These secondary structures assemble in to three different types of acidic (pI = 4.75 ~ 5.40, MW ~ 35,000 Da) and peptide chains and three basic (pI = 8.00 ~ 8.50, MW ~ 20,000 Da) ones [32, 33].

Raman spectroscopy is a technique that measures the inelastic scattering (Raman scattering) of a chemical. When a light beam is shined upon a sample, most of the reflected and scattered portion will retain the same frequency as the incoming beam. However, a relatively small fraction will be scattered at a lower frequency due to the energy consumed by the samples, a phenomenon known as the Raman scattering. The Raman and infrared spectroscopies are complimentary to each other, in that the strong bands in one method often show weak signals in the other.

Raman spectroscopy also shows several peaks designated as amide I, II, III, IV, etc. The amide I band (and less frequently the amide II or III band) can be subjected to the same FSD procedure to obtain separate peaks corresponding to different secondary structures. Raman spectroscopy provides more details on the protein structure compared to FTIR. Number of studies have been reported on the structural change in soy proteins upon amidation [34], deamidation [35], acetylation [36], heat processing [37], hydrolysis [38], and treatment with pulse electrical field [39].

Proteins bearing these residues (especially Trp) exhibit significant ultraviolet (UV) absorbance, peaking around 280 nm (absorption maxima for Trp). Determination of the optical density at this wavelength (OD_{280}) has been employed as a simple method for directly estimating the protein content; this is also the principle for many standard HPLC-based methods for analysing proteins. Trp in its free (nonprotein) form exhibits strong fluorescence with an excitation and emission wavelength of 280 nm and 348 nm, respectively. Free Tyr/Phe can also be excited at 274/257 nm and emit fluorescence at 303/282 nm, although their quantum yields are not as high. Magnitude of fluorescence declines as Trp is buried in a fully folded protein molecule. On the contrary, the denaturation of soy proteins exposes the hydrophobic Trp, Tyr, and Phe to the solvent, causing a red shift together with higher fluorescence intensity [40]. Providing qualitative information on the protein conformation, such studies demonstrate the potential of solving the binding affinity between the ligand and protein [41]. However, delicate correction must be taken to avoid artefacts arising from the inner filter effect [42].

Nuclear magnetic resonance (NMR) is a physical phenomenon in which nuclei in a magnetic field absorb and re-emit electromagnetic radiation. This energy is at a specific resonance frequency which depends on the strength of the magnetic field and the magnetic properties of the isotope of the atoms. The ^{13}C NMR spectrum was much less resolved as a result of the slow protein tumbling in solution and the presence of structured domains within the subunit pairs. The spectrum of the alkali-denatured protein (pH 12.0) was resolved satisfyingly, while the ones for the protein treated at pH 10 and 7 showed no observable difference. Mass spectrometry (MS) is an analytical chemistry technique that helps identify the amount and type of chemicals present in a sample by measuring the mass-to-charge ratio and abundance of gas-phase ions.

After being coated on the surface of the sample, the matrix is ionized and traverses in the electromagnetic field along with the samples, allowing them to be separated based on their mass-to-charge ratios. Electron paramagnetic resonance (EPR) or electron spin resonance (ESR) spectroscopy is an advanced technique for investigating the unpaired electrons of a molecule. Boatright *et al.* compared the contents of carbon-centred free radicals in soy proteins, sodium caseinate, and ovalbumin [43].

The result suggested a much higher content of free radicals in SPI compared to the other two. Small angle X-ray scattering (SAXS) is a technique that records the elastic scattering of X-ray by a molecule at very low angles (0.1° ~ 10°). SAXS is frequently exploited for characterizing the supramolecular structure of soy proteins. The data acquired from the SAXS study include radius of gyration R_g, scattering intensity at zero angle $I(0)$, and fractal dimension of the aggregates. The fractal dimension derived from the slope of SAXS curves at high angles is an index for particle compactness, with higher values indicating for more compact particles [44]. Wide angle X-ray diffraction (WAXS) is virtually the same technique as the SAXS, except that the working distance is shorter and the angles measured are therefore larger. Commercial and most lab-synthesized soy protein products are amorphous rather than crystalline. Therefore, they exhibit wide humps without a recognizable peak in a typical diffractogram.

Appearance of these characteristic peaks in the final product (e.g., a soy protein film) is a good indicator of the existence of the corresponding crystals. Following the same principle, it can be deduced that the newly formed sharp peaks in a polymer (or a polymeric blend) imply enhanced crystallization. Neutron scattering is the phenomenon that free neutrons by a molecule can refer to either the physical process or the experimental technique which uses this process for the investigation of materials. The neutron scattering is capable of distinguishing ^1H from ^2H; therefore, it is of great interest when the samples can be partly or fully deuterated as isotopic labels in the sample. The morphology of soy protein-based materials is a determinant factor on their chemical, mechanical, and functional properties. A thin layer of conductive material (e.g., graphite or gold) is usually sputtered onto the protein samples to improve image quality and avoid the charging effect [45].

The major advantages of SEM are the simplicity of sample preparation and the capability to create images with a certain depth. Transmission electron microscopy (TEM) gathers the electrons that pass through a thin layer of the sample. TEM is only sensitive to the surface of a sample, and thus it can provide detailed information on the surface features. It can be seamlessly integrated with energy dispersive X-ray (EDX) spectroscopy. Such technique is a convenient tool for determining the elemental composition of a certain spot on the SEM/TEM image, as well as an element distribution map of the whole image. Atomic force microscopy (AFM) is distinct from the other two techniques, in that it measures the distance and force directly by pressing a very sharp tip onto the surface of the sample. The unique advantage of AFM is the construction of three dimension images within a short time due to the mechanism of measurement. The shortcomings of this technique include small scanning area and the susceptibility to interference arising from nonlinearity and hysteresis [46].

Thermal studies provide important guidance on the thermal behavior of soy protein-based material. Three techniques have been applied for the investigation on soy proteins. Thermal gravimetric analysis (TGA) is a straight forward method that evaluates the weight loss of a material upon heating. Studies on the heat-induced disintegration of soy protein-based films [47], adhesives [48], fibres [49], hydrogels [50], plastics [51], and nanocomposites [52] using TGA have been extensively reported. Differential scanning calorimetry (DSC) measures the change in heat flow as a function of changing temperature.

Dynamic mechanical analysis (DMA) is mostly used for characterizing the viscoelastic properties of a polymer. Such test applies a varying frequency of stress or temperature to the material, leading to the variation in the complex moduli.

2.7 Conclusion

Soy proteins have emerged as versatile materials in science and technology. Herein, an overview of the structure-property relationship has been discussed regarding synthesis, structures, properties and applications of soy proteins. To meet the requirements of different applications both nanostructures and properties need to be tuned.

References

1. Petruccelli, S., Anon, M.C., Soy protein isolate components and their interactions. *J. Agric. Food. Chem.*, 43, 1762–1767, 1995.
2. Deak, N.A., Murphy, P.A., Johnson, L.A., Fractionating soybean storage proteins using Ca^{2+} and $NaHSO_3$. *J. Food Sci.*, 71, C413–C424, 2006.
3. Zhang, Z., Hua, Y., Urea-modified soy globulin proteins (7S and 11S): effect of wettability and secondary structure on adhesion. *J. Am. Oil Chem. Soc.*, 84, 853–857, 2007.
4. Vu Huu, T., Kazuo, S., Beta-conglycinin from soybean proteins. Isolation and immunological and physicochemical properties of the monomeric forms. *Biochimica et Biophysica Acta (BBA) - Protein Structure*, 490, 370–384, 1977.
5. Wolf, W.J., Soybean proteins. Their functional, chemical, and physical properties. *J. Agric. Food. Chem.*, 18, 969–976, 1970.
6. Kinsella, J., Functional properties of soy proteins. *J. Am. Oil Chem. Soc.*, 56, 242–258, 1979.
7. Nishinari, K., Fang, Y., Guo, S., Phillips, G.O., Soy proteins: A review on composition, aggregation and emulsification. *Food Hydrocolloid.*, 39, 301–318, 2014.
8. Khatib, K.A., Herald, T.J., Aramouni, F.M., MacRitchie, F., Schapaugh, W.T., Characterization and functional properties of soy β-conglycinin and glycinin of selected genotypes. *J. Food Sci.*, 67, 2923–2929, 2002.
9. Kumar, R., Choudhary, V., Mishra, S., Varma, I.K., Mattiason, B., Adhesives and plastics based on soy protein products. *Ind. Crops Prod.*, 16, 155–172, 2002.
10. Rickert, D.A., Meyer, M.A., Hu, J., Murphy, P.A., Effect of extraction pH and temperature on isoflavone and saponin partitioning and profile during soy protein isolate production. *J. Food Sci.*, 69, C623–C631, 2004.
11. Ali, F., Ippersiel, D., Lamarche, F., Mondor, M., Characterization of low-phytate soy protein isolates produced by membrane technologies. *Innov. Food Sci. Emerg. Technol.*, 11, 162–168, 2010.
12. Shallo, H.E., Rao, A., Ericson, A.P., Thomas, R.L., Preparation of soy protein concentrate by ultrafiltration. *J. Food Sci. -Chicago-*, 66, 242–246, 2001.
13. Nagano, T., Hirotsuka, M., Mori, H., Kohyama, K., Nishinari, K., Dynamic viscoelastic study on the gelation of 7S globulins from soybeans. *J. Agric. Food. Chem.*, 40, 941–944, 1992.
14. Deak, N., Murphy, P., Johnson, L., Characterization of fractionated soy proteins produced by a new simplified procedure. *J. Am. Oil Chem. Soc.*, 84, 137–149, 2007.
15. Kato, A., Osako, Y., Matsudomi, N., Kobayashi, K., Changes in the emulsifying and foaming properties of proteins during heat denaturation. *Agric. Biol. Chem.*, 47, 33–37, 1983.

16. Zhang, J., Liang, L., Tian, Z., Chen, L., Subirade, M., Preparation and *in vitro* evaluation of calcium-induced soy protein isolate nanoparticles and their formation mechanism study. *Food Chem.*, 133, 390–399, 2012.
17. Renkema, J.M.S., Gruppen, H., van Vliet, T., Influence of pH and ionic strength on heat-induced formation and rheological properties of soy protein gels in relation to denaturation and their protein compositions. *J. Agric. Food. Chem.*, 50, 6064–6071, 2002.
18. Teng, Z., Liu, C., Yang, X., Li, L., Tang, C., Jiang, Y., Fractionation of soybean globulins using Ca^{2+} and Mg^{2+}: a comparative analysis. *J. Am. Oil Chem. Soc.*, 86, 409–417, 2009.
19. Liu, M., Lee, D.-S., Damodaran, S., Emulsifying properties of acidic subunits of soy 11S globulin. *J. Agric. Food Chem.*, 47, 4970–4975, 1999.
20. Rickert, D.A., Johnson, L.A., Murphy, P.A., Functional properties of improved glycinin and β-conglycinin fractions. *J. Food Sci.*, 69, FCT303–FCT311, 2004.
21. Malhotra, A., Coupland, J.N., The effect of surfactants on the solubility, zeta potential, and viscosity of soy protein isolates. *Food Hydrocolloid.*,18, 101–108, 2004.
22. Tang, C.-H., Wu, H., Chen, Z., Yang, X.-Q., Formation and properties of glycinin-rich and β-conglycinin-rich soy protein isolate gels induced by microbial transglutaminase. *Food Res. Int.*, 39, 87–97, 2006.
23. Gutiérrez, J., McGinity, J.W., Influence of water soluble and insoluble plasticizers on the physical and mechanical properties of acrylic resin copolymers. *Int. J. Pharm.*, 103, 293–301, 1994.
24. Zhu, Y., Shah, N.H., Malick, A.W., Infeld, M.H., McGinity, J.W., Solid-state plasticization of an acrylic polymer with chlorpheniramine maleate and triethyl citrate. *Int. J. Pharm.*, 241, 301–310, 2002.
25. Gennadios, A., Brandenburg, A.H., Weller, C.L., Testin, R.F., Effect of pH on properties of wheat gluten and soy protein isolate films. *J. Agric. Food. Chem.*, 41, 1835–1839, 1993.
26. Hartman, G.H., Removal of phytate from soy protein. *J. Am. Oil Chem. Soc.*, 56, 731–735, 1979.
27. Brooks, J.R., Morr, C.V., Phytate removal from soy protein isolates using ion exchange processing treatments. *J. Food Sci.*, 47, 1280–1282, 1982.
28. Engelman, H.M., Alekel, D.L., Hanson, L.N., Kanthasamy, A.G., Reddy, M.B., Blood lipid and oxidative stress responses to soy protein with isoflavones and phytic acid in postmenopausal women. *Am. J. Clin. Nutr.*, 81, 590–596, 2005.
29. Huang, W., Sun, X., Adhesive properties of soy proteins modified by urea and guanidine hydrochloride. *J. Am. Oil Chem. Soc.*, 77, 101–104, 2000.
30. Susi, H., Byler, D.M., Resolution-enhanced fourier transform infrared spectroscopy of enzymes. *Methods Enzymol.*, 130, 290–311, 1986.

31. Byler, D.M., Susi, H., Examination of the secondary structure of proteins by deconvolved FTIR spectra. *Biopolymers*, 25, 469–487, 1986.
32. Mujoo, R., Trinh, D.T., Ng, P.K.W., Characterization of storage proteins in different soybean varieties and their relationship to tofu yield and texture. *Food Chem.*, 82, 265–273, 2003.
33. Catsimpoolas, N., Isolation of glycinin subunits by isoelectric focusing in urea-mercaptoethanol. *FEBS Lett.*, 4, 259–261, 1969.
34. Wong, H.-W., Phillips, D.L., Ma, C.-Y., Raman spectroscopic study of amidated food proteins. *Food Chem.*, 105, 784–792, 2007.
35. Wong, H.-W., Choi, S.-M., Phillips, D.L., Ma, C.-Y., Raman spectroscopic study of deamidated food proteins. *Food Chem.*, 113, 363–370, 2009.
36. Yu, Z., Ma, C.-Y., Yuen, S.-N., Phillips, D.L., Raman spectroscopic determination of extent of O-esterification in acetylated soy protein isolates. *Food Chem.*, 87, 477–481, 2004.
37. Herrero, A.M., Jiménez-Colmenero, F., Carmona, P., Elucidation of structural changes in soy protein isolate upon heating by Raman spectroscopy. *Int. J. Food Sci. Tech.*, 44, 711–717, 2009.
38. Phoon, P.Y., San Martin-Gonzalez, M.F., Narsimhan, G., Effect of hydrolysis of soy β-conglycinin on the oxidative stability of O/W emulsions. *Food Hydrocolloid.*, 35, 429–443, 2014.
39. Han, Z., Raman spectra study of soy protein isolate structure treated with pulsed electric fields. *Spectrosc. Spect. Anal.*, 30, 3236–3239, 2010.
40. Keerati-u-rai, M., Miriani, M., Iametti, S., Bonomi, F., Corredig, M., Structural changes of soy proteins at the oil-water interface studied by fluorescence spectroscopy. *Colloids Surf. B. Biointerfaces*, 93, 41–48, 2012.
41. Cogan, U., Kopelman, M., Mokady, S., Shinitzky, M., Binding affinities of retinol and related compounds to retinol binding proteins. *Eur. J. Biochem.*, 65, 71–78, 1976.
42. Van de Weert, M., Stella, L., Fluorescence quenching and ligand binding: a critical discussion of a popular methodology. *J. Mol. Struct.*, 998, 144–150, 2011.
43. Boatright, W.L., Jahan, M.S., Walters, B.M., Miller, A.F., Cui, D., Hustedt, E.J., Lei, Q., Carbon-centered radicals in isolated soy proteins. *J. Food Sci.*, 73, C222–C226, 2008.
44. Guo, J., Yang, X.-Q., He, X.-T., Wu, N.-N., Wang, J.-M., Gu, W., Zhang, Y.-Y., Limited aggregation behavior of β-conglycinin and its terminating effect on glycinin aggregation during heating at pH 7.0. *J. Agric. Food. Chem.*, 60, 3782–3791, 2012.
45. Teng, Z., Luo, Y., Wang, Q., Nanoparticles synthesized from soy protein: preparation, characterization, and application for nutraceutical encapsulation. *J. Agric. Food. Chem.*, 60, 2712–2720, 2012.
46. Lapshin, R.V., Analytical model for the approximation of hysteresis loop and its application to the scanning tunneling microscope. *Rev. Sci. Instrum.*, 66, 4718–4730, 1995.

47. Guo, G., Zhang, C., Du, Z., Zou, W., Li, H., Structure and properties of poly (vinyl alcohol)/soy protein isolate blend film fabricated through melt processing. *J. Polym. Environ.*, 23, 183–198, 2015.
48. Qu, P., Huang, H., Wu, G., Sun, E., Chang, Z., The effect of hydrolyzed soy protein isolate on the structure and biodegradability of urea-formaldehyde adhesives. *J. Adhes. Sci. Technol.*, 29, 502–517, 2015.
49. Sudha, T.B., Thanikaivelan, P., Ashokkumar, M., Chandrasekaran, B., Structural and thermal investigations of biomimetically grown casein-soy hybrid protein fibers. *Appl. Biochem. Biotechnol.*, 163, 247–257, 2011.
50. Liu, Y., Cui, Y., Yin, G., Luo, L., Preparation and properties of novel pH-sensitive soy protein/poly(sodium acrylate) interpenetrating polymer network hydrogels. *J. Biobased Mater. Bioenerg.*,3, 437–442, 2009.
51. Wang, H., Jiang, L., Fu, L., Properties of molded soy protein isolate plastics. *J. Appl. Polym. Sci.*, 106, 3716–3720, 2007.
52. Chen, P., Zhang, L., Interaction and properties of highly exfoliated soy protein/montmorillonite nanocomposites. *Biomacromolecules*, 7, 1700–1706, 2006.

3

Advances in Soy Protein-Based Nanocomposites

Huafeng Tian[1], Gaiping Guo[3] and Xiaogang Luo[2]*

[1]*School of Material and Mechanical Engineering, Beijing Technology and Business University, Beijing, China*
[2]*Key Laboratory for Green Chemical Process of Ministry of Education; Hubei Key Laboratory for Novel Reactor and Green Chemistry Technology; School of Chemical Engineering and Pharmacy, Wuhan Institute of Technology, Wuhan, China*
[3]*Department of Chemical Engineering, Tsinghua University, Beijing, China*

Abstract
As a kind of a by-product of soy oil industries, soy protein possesses good biodegradability, and is regarded as a viable alternative for petroleum-based polymeric products. Aiming to improve the physical properties, especially the mechanical properties and water resistance which limit their extensive use, incorporation of nanoparticle proves to be an effective way. The properties of the resulting nanocomposites are highly dependent on the processing methods, the nature of nanofillers, as well as the dispersion effect of the filler in the matrix. This chapter reviewed the fabrication methods, property-structure relationship, and application of soy protein nanocomposites. These biodegradable nanomaterials would find potential applications in edible films, packaging films, foams, biomedical materials, etc. For the environmental crisis caused by the extensive use of petroleum-based non-degradable materials, these eco-friendly biopolymers based products would find more and more applications in the future.

Keywords: Soy protein, nanocomposites, fabrication methods, structure-property relationship

**Corresponding author*: xgluo@wit.edu.cn, xgluo0310@hotmail.com

Visakh P. M and Olga Nazarenko (eds.) Soy Protein-Based Blends, Composites and Nanocomposites, (39–66) 2017

3.1 Introduction

To advance the sustainable development of human being, polymers from natural resources have attracted much attention for their biodegradable properties and low cost [1]. Protein-based "green" materials have become a research focus because of their high performance, low cost, and eco-friendly character [2]. Residual soy proteins, a by-product of the soy oil industry, are currently utilized in applications such as animal feed and food supplement. The raw commercial materials, soy protein isolate (SPI) containing more than 90% protein, soy protein concentrate (SPC) with about 70% protein, defatted soy flour (SF) containing about 50% protein, are commonly available.

Based on their ultra-centrifugal sedimentation rates, soy proteins can be sorted into four main categories: 2S, 7S, 11S, and 15S [3]. Soy proteins are composed of a mixture of albumins and globulins, 90% of which are storage proteins with globular structure, consisting mainly 7S (conglycinin) and 11S (glycinin) globulins. Soy protein contains 18 amino acids including those containing polar functional groups, such as carboxyl, amine and hydroxyl groups that are capable of chemically reacting and making soy protein easily modification [4]. However, the numerous polar functional groups also result in the water sensitivity, poor process ability and inferior mechanical strength of soy protein-based films, leading to the substantially limited application [5].

Temperature and pressure are two of the main parameters in soy protein processing in order to denature the protein, unfold globular structure, and permit interaction and entanglement between protein changes to modify material properties [6]. Therefore, intra- and intermolecular interactions will be significantly influenced by the processing temperature and pressure employed. Heating soy protein dispersions at temperatures above 70 °C permits unfolding of the globular structure, so that protein denatures, and forms new intra- and intermolecular bonding, such as hydrogen bonds or electrostatic or hydrophobic interactions. In the same way, at alkaline pH values (about 10), globulins also dissociate and the polypeptides irreversibly unfold. For glycerol-plasticized soy protein isolate films prepared by compression, the best mechanical properties were obtained when basic pHs were used due to a major unfolding of the protein which allowed the exposition of the polar groups to be able to interact with small glycerol molecules.

It is known that plasticizers with characteristics such as small size, high polarity, more than one polar group per molecule, generally impart great plasticizing effect on polymeric systems. Currently, biopolymer films are usually plasticized by hydroxyl compounds [7]. Glycerol has a high boiling point and good stability, and is regarded as one of the most efficient plasticizers for soy protein plastics [8]. Glycerol-plasticized soy protein possesses good processing properties and mechanical performance [9]. Some amide and hydroxylamine could also be used as soy protein plasticizers [10]. With the incorporation of small molecular plasticizers, the flexibility (especially elongation at break) of soy protein materials was highly improved, with the sacrifice of the strength and modulus [11].

Nanophase reinforcement in polymers and adhesives effectively improved their performance. An emerging class of hybrid materials represented by bionanocomposites, derived from combination of particles at the nanoscale with biopolymers matrix, has proven to be a promising option in improving mechanical and barrier properties of biopolymers [12]. The bionanocomposites consist of a biopolymer matrix reinforced with particles having at least one dimension in the nanometer range (1–100 nm) and exhibits much improved properties due to high aspect ratio and high surface area of the nanoparticles. When very low nanoparticle loading of 1–5 wt% was filled into matrices, mechanical, thermal, barrier properties, water resistance, and other properties of soy protein materials were improved [13]. For such composites, inorganic or organic nanoparticles have been introduced into soy protein matrix. The properties of nanocomposite materials depend not only on the properties of their individual parents but also on their morphology and interfacial characteristics. In this chapter, it is aimed to highlight the principal fabrication approaches, and property-structure relationship of soy protein base bionanocomposites.

3.2 Preparation Methods of Soy Protein Nanocomposites

3.2.1 Solution Casting

The soy protein-based nanocomposites could be fabricated though mixing the soy protein solution and nanoparticle dispersion, and then solution casting. The composite films could be obtained when the solvent evaporated. Water is the most used dispersion media for the hydrophilic nature of soy protein. However, for solution casting technique is energy costing and may bring out pollution, it is only possible to be used in the laboratory work. Also the cast film is often difficult to removal from the support when dried.

Soy protein films reinforced with starch nanocrystals (SNC) could be prepared by casting method [14]. The SNC synthesis was developed by acid hydrolysis of native cornstarch. SPI with glycerol plasticizer was dispersed in water and mixed with SNC dispersion. And then the mixed solution was poured into plastic Petri dishes and dried to obtain transparent nanocomposite films.

In addition of casting, the mixed solution could also be dried by lyophilization. For example, desired weighted SPI and chitin whiskers with the various contents were mixed and stirred to obtain a homogeneous dispersion. The dispersion was freeze-dried, and 30% glycerol was added. The resulting mixture was hot-pressed to obtain SPI/chitin whisker nanocomposite films [15].

3.2.2 Extrusion

Extrusion is one of the prime processes used in the plastics industry. As an efficient, simple, and fast preparation polymer processing technique, it is used to produce large numbers of identical items from high precision engineering components to disposable consumer goods, being very adequate to achieve a high-volume production [16]. Twin-screw extruders have considerably more heat exchange capability and higher production efficiency than single-screw extruders.

Soy protein is one of the few natural polymers that can be thermoplastically processed under the plasticization of small molecules [17]. Soy protein plastics without any additive have a brittle behavior, which makes processing difficult. Addition of plasticizers is an effective way to improve the flowability of soy protein melts and obtain flexible soy protein-based films. Thermal extrusion exposes the protein ingredients to high temperature, high pressure, and mechanical shear, which converts soy protein into a continuous plastic "melt," resulting in protein denaturation and solubility reduction [18].

Bionanocomposite films based on soy protein isolate and montmorillonite (MMT) were prepared using melt extrusion [19]. SPI (70–85%, dry basis), glycerol (15%, dry basis), and MMT (0–15%, dry basis) were mixed and left at room temperature for 2 h for hydration. The mixture was subsequently extruded in a twin-screw co-rotating extruder with as crew diameter of 25 mm and length to diameter ratio (L/D) of 20. The extrudate was dried and grounded in a grinder for further testing and film casting. The extrusion processing parameters (screw speed and barrel temperature distribution) could influence the mechanical properties of SPI/MMT films. There was a significant increase in tensile strength (TS) of the films with increasing screw speed and barrel temperature. Improved TS at higher screw speed can be attributed to better dispersion of MMT. Higher shear rate, corresponding to higher screw speed, can break bigger agglomerates of MMT into smaller aggregates. Improved TS at higher barrel temperature can be attributed to a decrease in viscosity with increase in temperature. The stress required to break MMT aggregates is reduced as the viscosity decreases. An increase in barrel temperature can also increase the extent of protein denaturation. Denatured proteins will have more chains available for network formation, resulting in improved TS.

3.3 Properties of Thermoplastic Soy Protein Nanocomposites

Nanoparticles have been commonly used to improve polymer properties. It has been shown that a small loading of nanoparticles can significantly improve the mechanical and thermal properties of the polymers. Some other properties, such as barrier properties, thermal stability, are also closely related to the content and morphology of the filler.

3.3.1 Mechanical Properties

Neat protein-based plastics usually exhibited fragile mechanical properties. When plasticized with small molecules, soy protein behaves like thermoplastic materials with good flexibility, but low mechanical strength and poor water resistance [20]. Incorporation of fillers constitutes an effective way to improve the strength of soy protein plastics. The interaction and compatibility between the components played a key role in reinforcing. The stress of the matrix would be transferred to the filler through interface on the assumption that fine compatibility exist between the filler and matrix. Phthalic anhydride modified soy protein (PAS)/glycerol plasticized soy protein (GPS) composite films were fabricated by using extrusion and compression-moulding [21]. Figure 3.1 shows the synthetic route for PAS. Phthalic anhydride could react with $-NH_2$ groups on soy protein, resulting

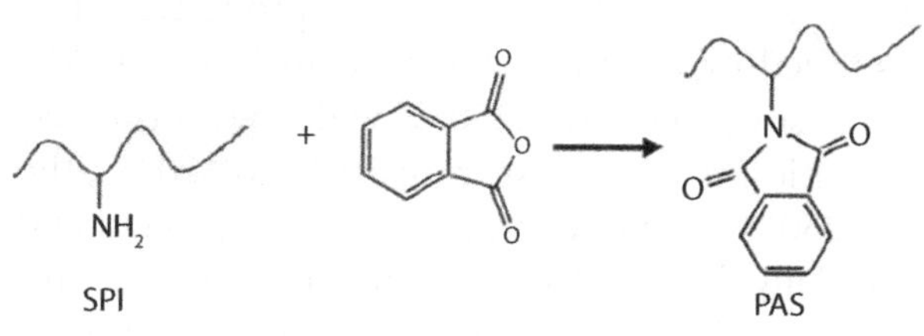

Figure 3.1 Synthetic route for *N*-phthaloyl soy protein.

in the decrease of the polar groups and the loss of plasticibility after modification. Therefore, the modified soy protein could be used as a filler to reinforce the GPS matrix. The similar chemical structure of PAS and GPS led to compatibility of the two components, resulting in high transparency and enhanced mechanical properties of the composites. PAS provided obvious reinforcing effect on GPS, resulting in higher TS and Young's modulus, however, the elongation at break decreased.

3.3.2 Water Resistance Properties

Water sensitivity of soy protein materials is one of the main drawbacks which prevent its wide usage. The polar groups on soy protein macromolecules endow the material hydrophilicity. When polyol plasticizers were added, the water sensitivity was further increased, leading to the variable mechanical properties under different environment relative humidity (RH). Incorporating nanofillers is an effective way to improve the water resistance of soy protein. The interaction between the filler and matrix could prevent the swelling of protein matrix in high RH, leading to the improvement of the water resistance of the resulting nanocomposites. For previously discussed PAS/GPS composite films [21], with increasing filler content, the water uptake decreased linearly. PAS would exhibit hydrophobic behavior because of the lowered polar groups, resulting in the improved water resistance. In addition, because of the strong interactions between GPS matrix and PAS filler, the moisture was blocked from the matrix and the overall water uptake was decreased.

3.3.3 Thermal Stability

Boron nitride (BN) has many unique properties like thermal conductivity, high melting and decomposition temperature (2700–3000 °C), very low density (2.25 g/cm^3), high thermal stability, oxidation resistance, chemically inertness, limited surface activity, nonwet ability and lubricating effect. Soy/BN nanocomposites were prepared by low-cost green technique with water as the solvent [22]. The thermal properties of the nanocomposites were studied by thermogravimetric analysis (TGA). The thermal decomposition of soy protein was 250 °C which was occurred with a maximum decomposition to water loss from 30 °C to 120 °C, the second step from 120 °C to 250 °C due to degradation of soy protein and third to the oxidation of partially degraded soy protein and then charring. Soy protein decomposed completely at about 535 °C whereas a significant amount of residue was observed in the case of soy/BN nanocomposites. The residue left after weight loss at 800 °C was about 40% more in the case of soy/BN (10 wt%) nanocomposite. Hence, the thermal stability soy/BN nanocomposite was more than that of virgin matrix due to incorporation of thermally stable nano-BN.

The fire hazard or flammability of materials is a matter of major interest to many public and private agencies. Halloysite nanotube (HNT) is a unique form of aluminosilicate clay with a hollow tube structure. HNT nanoparticles are available from a variety of mineral deposits but show some additional promise as flame retardants because they do not typically require an organic treatment to be dispersed in polymer, and they have shown the ability to lower mass loss rate during burning [23]. Nakamura *et al.* investigated the effects of HNT on the flammability properties of green composites from jute fabric and SPC [24]. The composites tested show behavior typical of charring polymeric materials when burned. SPC has lower flammability (heat release capacity) than petrochemical-based resins, such as epoxies and vinyl esters. Glycerol has a negative effect on flammability (increased heat release), whereas the addition of HNT has a positive effect on flammability (decreased heat release) at the expense of an earlier time to ignition. Incorporating 5% mass fraction of HNT is found to reduce the composite flammability, while having no negative impact on the mechanical properties.

The limiting oxygen index (LOI) is a method for evaluation of the flammability of materials. LOI is defined as the minimum concentration of oxygen in an oxygen–nitrogen mixture, required to sustain burning of a vertically mounted specimen. Hence, higher LOI values represent better flame retardancy and a smaller LOI represents a more flammable material. As air contains 21% O_2, materials with an LOI below 21% are combustible, whereas those with an LOI above 21% are self-extinguishing [25].

Cellulose whiskers (CWs) in combination with nanoclay were used as reinforcing agents for the preparation of green nanocomposites based on jute fabric and glutaraldehyde (GA) cross-linked soy flour [26]. All of the samples produced small localized flames and that nanoclay-filled samples generated higher char than those without nanoclay. Also the LOI value increases with increasing percentage of CWs in the composites. CWs form cross-links with SF, jute, and GA and thus restrict the accessibility of oxygen for the production of degradable components from the composites, hence resulting in higher LOI values. Upon addition of nanoclay into the CW-incorporated composites, the LOI value is further enhanced. The incorporation of nanoclay produces a silicate char on the surface and improves the flame resistance properties. The silicate-rich surface has better barrier properties to heat and oxygen transport, resulting in a delay of the ignition of the composite delays.

3.3.4 Biodegradability

Renewable polymers such as cellulose, starch, polysaccharides, and proteins, may be decomposed into CO_2, water, and organic molecules by microorganisms in the soil. Biodegradability of the samples could be studied by evaluating weight loss of the blends over time in soil environment. Samples were weighed and then buried in soil. The buried samples were dugout for a certain time, washed in distilled water, dried and weighed before returning them to soil. Also the degraded surface could be observed by SEM.

The biodegradation behaviors of maleated polycaprolactone/SPI composites reinforced with organoclay were evaluated by soil burial test [27]. Composites containing higher percentage of soy protein degraded rapidly in the initial 8 weeks and a gradual decrease of weight occurred during the next 8 weeks. Further, the time of degradation decreased with increasing organoclay content. Chen *et al.* prepared a series of soy protein/alumina hydrate nanocomposites via *in situ* reaction between aluminum chloride and ammonia in aqueous media [28]. The alumina hydrate particles with a dimension of about 10 ~ 50 nm were homogeneously dispersed in the soy protein matrices when the $AlCl_3$ addition was lower than 8 wt%. The biodegradability of the composites was investigated by C. olivaceum, T. Viride, and A. oryzae in culture medium. The SEM images of the degraded surfaces are shown in Figure 3.2. The surfaces of all plastic sheets exhibit coarse morphologies. After washing with 75% alcohol, there is still residual mycelium on some of the surfaces. For the plastics without filler, extent of degradation for each of three mildews is fairly high, and especially, those plastics sheets after seven-day exposure to T. viride or A. oryzae have been totally destructed into small fragments as shown in Figures 3.2 (b1) and (c1). For the composite plastics with high AlOH contents, relatively smooth surfaces are exhibited comparing with neat soy protein plastics (SA-0). The organic components of the protein plastics could be transformed into biomass, and promote the growth of mycelium. However, the speed of biodegradation process is slowed down because of the confinement effect and the alkaline nature of the AlOH nanoparicles.

3.3.5 Barrier Properties

Barrier properties of oxygen and carbon dioxide are essential for food packaging materials. Protection against oxygen is one determining factor that guarantees the maximum shelf life of food products [29]. Among these biopolymers, soy protein films are effective barriers to the passage

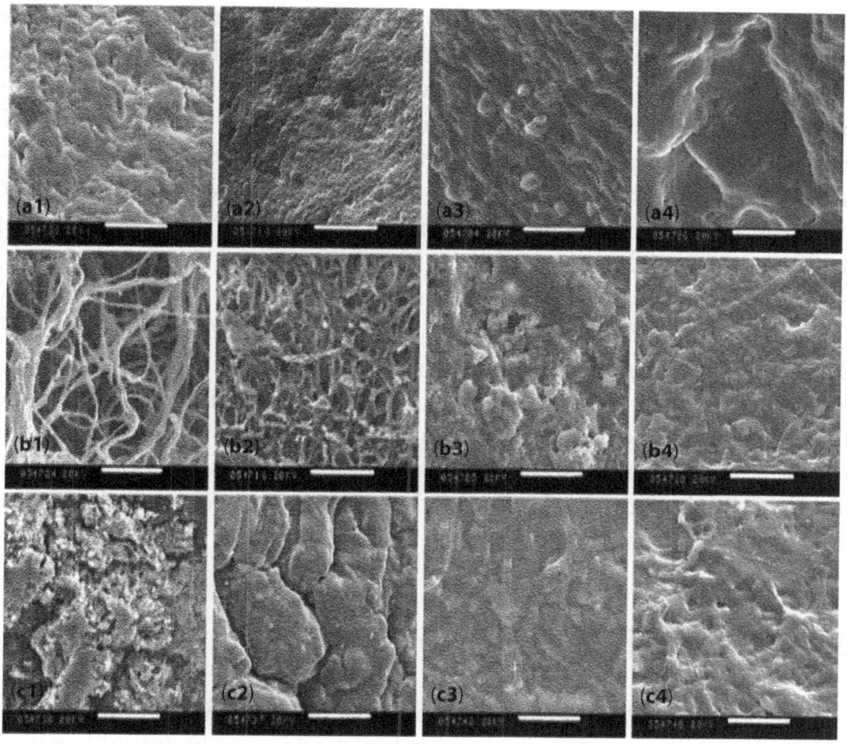

Figure 3.2 SEM images of the surfaces of SA-0 (a1, b1, and c1), SA-4 (a2, b2, and c2), SA-8 (a3, b3, and c3) and SA-20 (a4, b4, and c4) after seven-day exposure to three funguses at 37 °C. The number after SA means the content of $AlCl_3$. The character "a" in the labels means the plastic sheets were degraded by Chaetomium olivaceum, "b" means Trichoderma viride, and "c" means Aspergillus oryzae. The scale bar inside represents 30 μm.

of lipid, oxygen, and carbon dioxide. However, the inherent hydrophilicity of proteins and the substantial amount of plasticizer added in the film perform poorly in moisture barrier and mechanical properties as packaging material. Formulations with higher plasticizer content had lower barrier properties [30]. Nanoreinforcement of polymers to prepare nanocomposites has already proven to be an effective way to improve these properties concurrently [31]. The clay layers generate a tortuous pathway through which the permeable elements have much greater difficulty in penetrating the nanocomposite [32]. Lee *et al.* introduced MMT into glycerol plasticized soy protein composites and assessed the water vapor permeability (WVP) and oxygen permeability (OP) [33]. WVP decreased significantly as MMT content increased in SPI films. Permeability is controlled by diffusion and sorption. The reduction of

WVP can be attributed to a change in insolubility rather than to a change in diffusivity because of the lower availability of the hydrophilicsite for water vapor by establishment of hydrophilic interactions between proteins and nanoclays. Nanocomposite SPI films containing 3% MMT exhibited considerably reduced O_2 permeability, whereas films containing 15% MMT exhibited increased OP. This could be attributed to the superior level of organoclay intercalation achieved at Na-MMT levels of 3–12% compared with 15% MMT.

Carbon nanoparticles/soy protein isolate biofilms were prepared by casting method and the water barrier properties were evaluated [34]. WVP decreased significantly as the filler content increased. The hydrogen bond between filler and protein matrix lowered the availability of hydrophilic sites for water vapor, which was dominant in neat soy protein films. In addition, the formation of the compact and tough cross-sections reduced the interstitial spaces in the film and consequently decreased the rate of diffusion of water molecules through the films.

3.4 Protein-Based Nanocomposites

For the commonly used plasticizer is susceptible to water and environmental humidity, soy protein thermoplastics possess a hydrophilic nature and are susceptible to water and environmental humidity, leading to poor water resistance and mechanical properties of the obtained plastics [35]. One way for development of the new soy protein materials with improved properties is the reinforcement of soy protein with nanofillers, including inorganic fillers such as layer silicates, carbon nanotubes, and organic fillers, such as whiskers, nanofibrillar cellulose, etc. The nanoparticle dispersion is critical to obtain enhanced physical, mechanical, and thermal properties.

3.4.1 Soy Protein/Inorganic Filler Nanocomposites

Polymer/inorganic nanocomposites have a strong record in materials science because of the markedly improved physical and mechanical properties [36]. Silica is a kind of typical filler used in polymeric compounds as an effective reinforcing agent [37]. Nanocomposites were developed by *in situ* fabricating silica in soy protein isolate matrix with glycerol as plasticizer [38]. It was found that the silica particles synthesized in the protein environment with sodium silicate as a precursor, exhibited a uniform distribution in the protein matrix. Pronounced reinforcing effect was obtained with the dramatically enhanced TS and Young's modulus of the nanocomposite sheets. Furthermore, the thermal stability was improved with the temperatures at the maximum rate of weight loss determined by the peak of the DTG curves corresponding to glycerol release and protein degradation largely enhanced by adding silica. In addition of *in situ* synthesis, soy protein/silica nanocomposites could be fabricated through compounding nano-SiO_2 particles into soy protein isolate matrix [39]. The strength, modulus, and elongation were simultaneously enhanced when the nano-SiO_2 content was lower than 8 wt%.

Polyhedral oligomeric silsesquioxanes (POSS) is a three-dimensional structurally cage-like molecule. To enhance the mechanical and water-resistant properties of soy protein isolate based films, hydrophobic Trisilanolphenyl polyhedral oligomeric silsesquioxanes were incorporated to modify the SPI films [40]. SPI contains more than 90% proteins, counting 7.7% arginine (Arg), 6.8% lysine (Lys), and 2.5% cysteine (Cys) that has abundant of $-NH_2$ groups [41]. As shown in Figure 3.3, 3-Glycidoxypropyltrimethpxysilane (GPTMS) serviced as cross-linking agents between SPI and POSS. The $-NH_2$ groups on the SPI and the epoxy groups on the GPTMS can react with each other through the ring opening polymerization. POSS particles are clearly observed in the film without GPTMS (SPI/POSS). However, the film (SPI/GPTMS/POSS) became uniform after adding GPTMS. The ATR FTIR spectrum of SPI/GPTMS/POSS film showed the POSS peaks, but SPI/POSS film did not, indicating that POSS was not dispersed and reacted with SPI in the SPI/POSS film. Therefore, the mechanical properties and water resistance of SPI films were significantly improved by incorporating GPTMS and POSS.

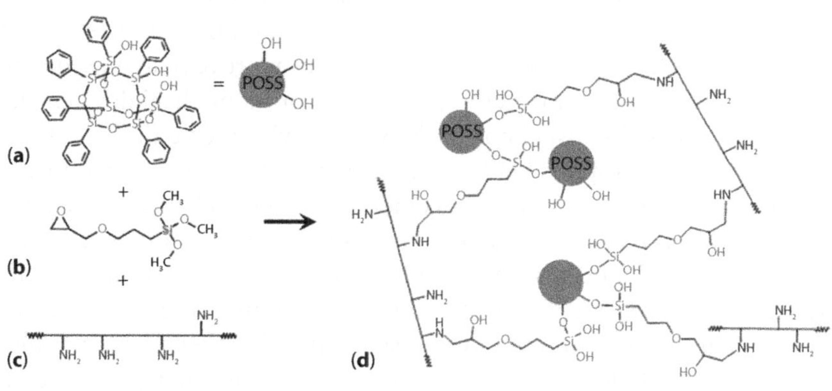

Figure 3.3 Illustration of the cross-linking reactions for SPI/GPTMS/POSS (d) from SPI (c), GPTMS (b) and POSS (a).

For the high stiffness and strength, as well as its excellent functional properties including biocompatibility, osteoconductivity, nontoxicity, and bioactivity, hydroxyapatite (HA) is a very promising reinforcing material for various polymers to fabricate composites for bio applications such as tissue scaffolds and bone regenerations [42]. Compared with commercially available soy protein which is very inexpensive, nanoscale HA is expensive due to the use of high purity materials and reagents used to produce it. Recently, a kind of nanoscale HA from eggshell waste was successfully synthesized by energy-efficient microwave-assisted wet chemical precipitation and then incorporated in soy-based resin by ensuring uniform dispersion by a polyelectrolyte coating. The coated nanoscale HA acted as effective filler for soy protein-based biopolymer in terms of improvement in functional properties of the nanocomposites. Significant improvements in tensile modulus and strength were achieved owing to the inclusion of uniformly dispersed HA in SPI sheets [43].

Besides forming HA, the eggshell itself could be used in the reinforcing of polymers. Rahman *et al.* synthesized a kind of eggshell nanopowder (ESNP) via size reduction of chicken eggshells using ball milling followed by a sonochemical process and prepared the eco-friendly hybrid nanocomposite from soy protein and eggshell ESNP [44]. The size distribution of the nanopowder was in the range of 20–40 nm. Due to its high crystallinity, surface area, thermal degradation temperature, and stiffness, eggshell nanopowder has been used successfully as nanofiller in bio-based resin to improve the functional properties of soy protein-based green composites. Significant improvements in Young's modulus, TS, and thermal stability of SPI nanocomposite sheets were achieved as a result of incorporation of well dispersed ESNP.

Clay is a green alternative to current practices and reduces flammability and permeability to liquids and gases in a wide range of plastics. The U.S. Food and Drug Administration has already approves clay-based composites for use as foods, medicines, beverages, or biomedical devices [45]. As is well known, the dispersion scale of the phyllosilicate in the polymer matrix is strongly responsible for the efficiency of the clay as reinforcing filler. There are four possible arrangements of layered clays dispersed in a polymer matrix–phase separated or immiscible (microcomposite), intercalated, exfoliated, and disordered intercalated (partially exfoliated). The intercalation is the state in which polymer chains are present between the clay layers, resulting in a multilayered structure with alternating polymer/inorganic layers. Exfoliation is a state in which the silicate layers are completely separated and dispersed in a continuous polymer matrix. The dispersion of MMT in a polymeric matrix depends on the process used in the preparation of the nanocomposite, the nature of both the polymer and the clay, and finally the interaction between those two integral components. Accordingly, the process and mechanism of the highly exfoliated polymer/clay nanocomposites have been a focus in the fields of fundamental and application research [46, 47]. Nanocomposites can be obtained by several methods which include *in situ* polymerization, solution exfoliation, and melt intercalation. The most common class of materials used as nanoparticles is layered clay minerals such as MMT. MMT has a very high elastic modulus (178 GPa) as compared to most biopolymers, which enables MMT to improve mechanical properties of biopolymers by carrying a significant portion of the applied stress. Various kinds of soy protein/MMT nanocomposites have been reported in previous literatures. Chen *et al.* prepared the biodegradable SPI/MMT plastics with highly exfoliated and intercalated structures via a solution intercalation process in neutral aqueous medium [48]. The structures of both the SPI/MMT nanocomposites and the plastics were strongly depended on the MMT content. When the MMT content was lower than 12 wt%, the MMT was highly exfoliated into single layers with a thickness of approximately 1–2 nm, whereas the intercalation structure predominated when the MMT content was higher than 12 wt%. The electrostatic surface potential calculation revealed that the heterogeneous distribution of the surface positive charges provided the possibility for negatively charged soy protein to intercalate and exfoliate MMT. Two kinds of interactions existed in this protein/MMT system, that is, the surface electrostatic interaction between the positive-charge-rich domains of soy protein and the negatively charged MMT layers as well as the hydrogen bonding between the –NH and Si–O groups. Such two interactions were beneficial to the intercalation and delamination of the MMT layers in the soy protein matrixes. Accordingly, the mechanical strength and thermo-stability of the SPI/MMT plastics were significantly improved as a result of the fine dispersion of the MMT layers and the strong restriction effects on the interfaces.

Jin revealed that surface-coating of MMT with a sufficient amount of SP at an appropriate acidity could enlarge the basal spacing of MMT platelets corresponding to highly intercalated and/or exfoliated layers [49]. When the protein-coated MMT is dispersed in a biopolymer matrix, functionalities of bionanocomposite systems can be strengthened by various approaches, for example, cross-linking protein molecules on MMT with the matrix biopolymer. They further prepared nanocomposite hydrogels with SP-coated MMT as nanofillers and SP as the matrix biopolymer [50]. The layered MMT nanoclay was intercalated by surface-coating with soy protein before mixing with 6% w/v SP for cross-linking by microbial transglutaminase (mTGase). The integration of surface-coating and mTGase cross-linking is promising to improve properties of the nanocomposite system.

Another kind of layered silicate, rectorite (REC), was also reported to be introduced into soy protein matrix [51]. The REC is composed of a regular (1:1) stacking of mica-like layers and MMT-like layers. Compared with the MMT interlayer spacing of about 1.2 nm, its interlayer spacing can reach about 2.4 nm. Thus, the REC was expected to allow soy protein molecules penetrating more availably, namely the REC lamellae are exfoliated more easily. The exfoliated REC lamellae resulted in enhancement of strength and modulus of SPI plastics.

Carbon nanotubes (CNs) are considered ideal reinforcing fillers for polymer matrices to achieve high performance and special functions, which can be attributed to their nanoscale, high flexibility, high aspect ratio, low mass density and, more importantly, their extraordinary mechanical strength and high electrical and thermal conductivity [52]. Zheng *et al.* reported the nanocomposite sheets by compounding MWNTs of various sizes into SPI matrix through solution mixing and then compression-molding method [53]. Improved mechanical performance and higher water resistance were obtained depending on MWNT size and content. In particular, a nanocomposite containing 0.25 wt.% of 10–15 nm MWNT was simultaneously reinforced and toughened. The improvement of mechanical properties was mainly attributed to interactions at the interface of SPI chain wrapped MWNT and the interface between penetrating SPI chains and the internal wall of MWNTs, as well as the associations with the SPI matrix mediated by protruding segments from the channel inside.

As a cheap and nontoxic filler, titanium dioxide (TiO_2) nanoparticles have been widely used in food packaging material, for they would provide protection against food borne microorganisms as well as odor, staining deterioration, and allergens on the presence of radiation of relatively low wavelength near the ultraviolet region [54, 55]. Nanocrystalline TiO_2 particles coated with soy protein isolate were fabricated in aqueous solution at near room temperature [56]. The film demonstrated the bactericidal efficiency of composite film against *Escherichia coli* and *Staphylococcus aureus*. When exposed to 2 h irradiation of UV light at 365 nm, the bactericidal activity of SPI films containing TiO_2 nanoparticles (2.00 g/150 mL) against gram-negative *E. coli* and gram-positive *S. aureus* can reach 71.01% and 88.49%, respectively. Therefore, the nanocomposite films with TiO_2 implanted have obviously better bactericidal efficiency. However, some aggregation and coalescence occurred with increasing TiO_2 content. Wang *et al.* prepared SPI/nano-TiO_2 films by solution casting and modified by ultrasonic/microwave assisted treatment (UMAT) [47]. Incorporation of nano-TiO_2 significantly enhanced films' mechanical properties and barrier properties, because of the intermolecular force between nano-TiO_2 and SPI. UMAT could promote the integration of fillers in SPI matrix and form closer intermolecular force between nano-TiO_2 and SPI, leading to a more compact and dense structure of films, and resulting in obviously improved TS and reduced WVP and OP.

In addition of TiO_2, silver nanoparticles are known to exhibit much more effective antimicrobial activity toward many kinds of bacteria, fungi and viruses compared with traditional bulk silver due to their unique surface effect, volume effect, and quantum size effect [57]. Zhao *et al.* embedded antimicrobial silver nanoparticles (AgNPs) into SPI film by *in situ* synthesized from SPI/$AgNO_3$ solution, taking the advantage of the reducibility of tyrosine residue in SPI, which accounts for approximately 4% in total protein content [58]. The AgNPs with the particle size of about 7 nm were homogeneously distributed without obvious aggregation. The antimicrobial test showed that the films had the excellent antimicrobial properties for killing both gram-positive bacteria (*S. aureus*) and gram-negative bacteria (*E. coli*) efficiently, suggesting that the addition of AgNPs has the effect to prevent bacteria growth on the SPI-based materials, and thus prolong their useful life, which endow them applications in antimicrobial food packages or biomedical fields.

3.4.2 Soy Protein/Organic Filler Nanocomposites

The organic fillers used in soy protein reinforcement are usually from biodegradable biopolymers, including starch, chitin, cellulose, etc. These naturally occurring materials forming in nature during the life cycles of green plants, animals, bacteria, and fungi are attracting more and more attentions for producing polymer matrix composites for their eco-friendly nature.

Starch is a kind of abundant recourses on the earth. Starch nanoparticles (SNs) were prepared by precipitating from gelatinized starch solution by ethanol, and further modified by citric acid to obtain citric acid-modified starch nanoparticles (CSN) [59]. The modified starch nanoparticles could not be gelatinized even after heating at 100 °C for 30 min as shown in Figure 3.4. When being incorporated as fillers in glycerol-plasticized soy protein plastics, the nanoparticles exhibited a homogeneous distribution in the matrix. CSN fillers enhanced the storage modulus of soy protein plastics and increased the glass transition temperatures due to the hydrogen bonding interactions between soy protein and the CSN filler. Also the

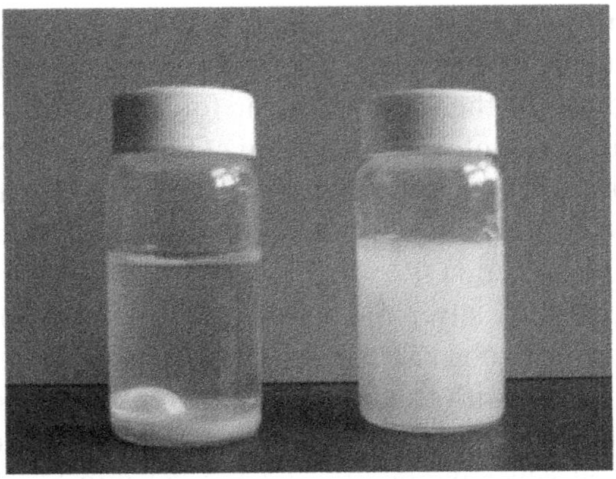

Figure 3.4 Citric acid-modified starch nanoparticles (left) and unmodifiedstarch (right) (5%, w/v in water) after heating a 100 °C.

mechanical properties of the resulting nanocomposites were improved. These protein nanocomposites will have potential applications in agriculture and packaging fields such as edible films, food packaging, and one-off packaging.

The advantages of natural polysaccharide whisker or nanocrystal are their low cost, low density, high specific strength and modulus, low energy consumption, renewable character, and biodegradability associated with the highly specific properties of nanoparticles [60]. Pea starch nanocrystals (StNs) were incorporated into SPI matrix to produce a class of full-biodegradable nanocomposites [61]. The nanocrystals present the originality to have a platelet-like morphology with a length of 60–150 nm and a width of 15–30 nm. The rigid StN showed a prominent reinforcing function, depending upon uniform dispersion and strong interfacial interaction between the StN filler and the SPI matrix. The nanocomposites containing 2 wt% showed the highest strength and Young's modulus, which was enhanced by ca. 50% and 200%, respectively, in contrast to neat SPI sheet. The water uptakes of the nanocomposites had no obvious decrease probably because of the hydrophilic nature of starch nanocrystals.

Cellulose nanocrystal (CNC), one of the most promising renewable nanomaterials, could be obtained from natural fibers, such as wood, cotton, etc. Because of its unique rod-like structure, high aspect ratio, high degree of crystallinity, and a large amount of hydroxyl groups on its surface, CNC has been widely used in polymer composites enforcement. A colloidal suspension of cellulose whiskers was introduced to SPI plastics, the results showed that both the TS and Young's modulus of the

SPI/cellulose whisker composites greatly increased [62]. The increased properties were mainly due to the intramolecular hydrogen bonds between matrix and the fillers. More announced composite performance could be achieved if strong, chemical bonding between the cellulose surfaces and matrix were successfully facilitated. Zhang *et al.* reported a novel method to attach functional groups to the CNC surfaces using surface modification [63]. CNC was firstly modified with a silane coupling agent for introducing reactive amino groups. Then the ethyleneglycol diglycidyl ether (EGDE) was used as a cross-linking agent in effort to improve the performance of the resultant modified-SPI films (Figure 3.5). As a result, the superficial reactive amino groups of the modified CNC(MCNC) facilitated interaction among SPI, MCNC, and EGDE and formed a larger network structure, and the interaction between the modified CNC and SPI matrix was enhanced by forming the chemical-cross-linking network through the reaction between amino and epoxy groups, thus improving the mechanical properties of the films. It was also accounted for the improvement of the thermal stabilities and water resistance of the MCNC-modified SPI films.

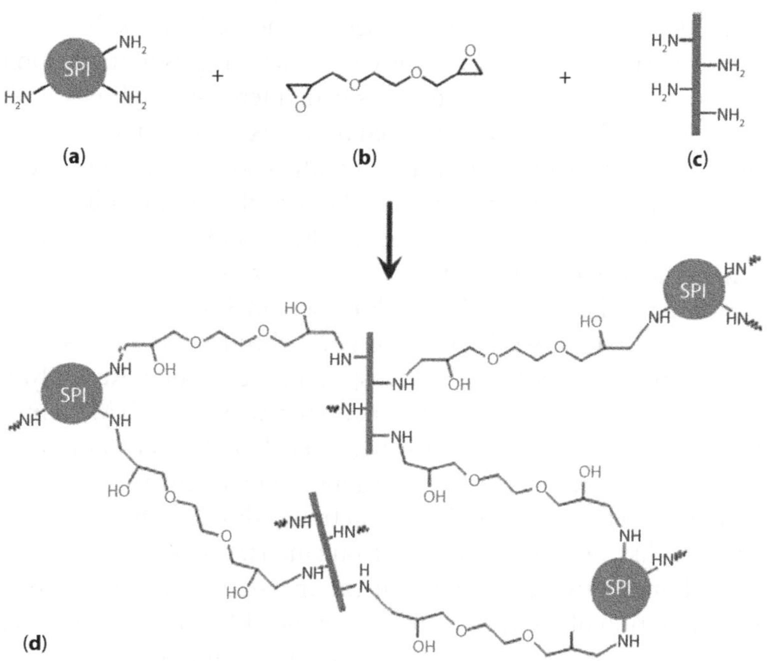

Figure 3.5 Reaction among SPI, MCNC, and EGDE (a) SPI, (b) EGDE, (c) MCNC, and (d) SPI-based films cross-linking networks.

Micro-fibrillated cellulose (MFC) obtained by mechanical shearing of cellulose fibers has high aspect ratio and high tensile properties and has been used for fabrication of composites with excellent mechanical properties [64]. Micro/nano-sized bamboo fibrils (MBFs) and a modified soy protein resin were used to fabricate environmentally friendly composites [65]. With the incorporation of MBF, the fracture stress and Young's modulus of the soy protein concentrate increased significantly. The addition of MBF, however, did not show significant decrease in the fracture strain of the specimens. As a result, the toughness of the MBF reinforced SPC increased. When cross-linked using a silane, (3-isocyanatopropyl) triethoxysilane (ITES), the fracture toughness was further enhanced.

Because of their distinctive properties such as low density, high surface area, and low thermal conductivity, aero gels can be used in the design of thermal insulators, porous catalysts for chemical processes and porous ceramics for filtration and separation. Organic aerogels based on soy proteins and nanofibrillar cellulose (NFC) are developed from precursor aqueous dispersions and a facile method conducive of channel- and defect-free systems after cooling and freeze-drying cycles [66]. The SP aerogels presented an off-white color and were brittle, while those obtained from pure NFC were brighter and rigid (Figure 3.6). The pores of SP aerogels seem to be more interconnected and have loosely-defined cell edges. NFC loading drives the internal morphology of the composite aerogels to transition from network- to fibrillar-like, with high density of interconnected cells. The brittleness of SP-based aerogels disappeared or was extensively reduced by addition of NFC loading. Composite aerogels with SP loadings as high as ca. 70% display a compression modulus of 4.4 MPa very close to that obtained from pure NFC aerogels. The potential functionalities of the SP–NFC green composite materials can benefit from the reduced material cost and the chemical features brought about the amino acids present in SPs.

Electrospun nanofiber is an emerging fibrousnano material generated from simple but versatile electrospinning technique. Recently, Chen reported the fabrication of cellulose nanofibrous mats (CNM) reinforced soy protein isolate composite with high visible light transmittance [67]. Because of the ultrafine diameter and super high length-to-diameter ratio of nanofibers, cellulose nanofibers penetrated through the SPI matrix to form an IPN-like network, in which strong interfacial interactions through hydrogen bonding were presented at the massive interfaces of fiber/SPI. The incorporation of 20 wt% nanofibers in the SPI matrix resulted in great improvement of mechanical strength and Young's modulus by respectively 13 and 6 times more than neat SPI film. More interestingly, this composite was translucent with light transmittance of over 75% at 700 nm. Such

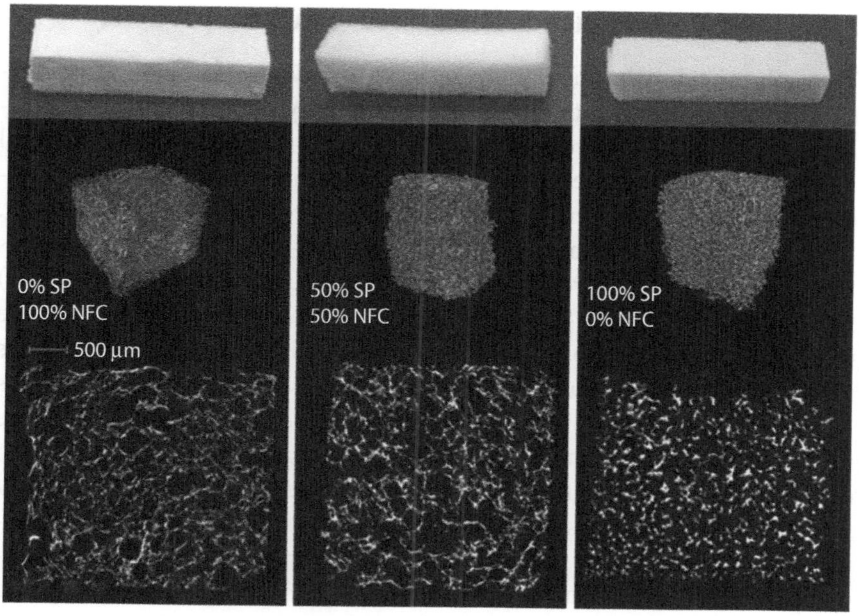

Figure 3.6 Upper section SP–NFC composite aerogels of a few selected compositions: 100% NFC, 50:50 SP–NFC and 100% SP. Lower section tomography images of cubic sections of the respective aerogel.

high light transmittance of cellulose nanofiber reinforced SPI compositeis closely associated with the strong fiber/SPI interfacial adhesion and the ultrafine fiber diameter.

Cyclodextrins (CDs) can thread the polymer chains to produce a rigid and cylindrical supramolecular polyrotaxanes (PRs). The CD-based inclusion was used to create stretching conformation of polymer segments in bulk material and usually removed by water washing or enzymatic degradation after changing the conformation of polymer chain and controlling polymorphic crystalline [68]. The self-assembled rigid supramolecular nanoplatelets (SN) from Pluronic polymers with various lengths of polyethylene oxide (PEO) and β-cyclodextrin have reinforced the soy protein isolate-based biodegradable plastics in terms of strength and modulus but at the expense of elongation [69]. A small amount of nanoplatelets was able to disperse uniformly into SPI matrix to produce optimal strength and modulus. The nanoplatelets with longest free PEO segments produced highest strength with less expense of elongation by virtue of enhanced association with SPI matrix mediated by PEO chains. Meanwhile, the water resistance was also enhanced.

3.4.3 Soy Protein Blends Reinforced with Nanoparticles

To resolve the drawbacks of water sensitivity and poor mechanical properties, blending SPI with other biodegradable polymers such as polycaprolactone (PCL) [70], poly(lactic acid) [71], poly(vinyl alcohol) [72], natural rubber [73], etc., thus becomes a way to enlarge its applications. The properties of the blend materials could further improved by nanoreinforcing. Sasmal *et al.* prepared a kind of biobased, eco-friendly nanocomposites from maleated polycaprolactone/soy protein isolate blend (50/50 wt/wt) reinforced with organo-modified clay by melt compounding [74]. The intercalated nanocomposite is formed and the silicate layers of the clay are uniformly dispersed at a nanometer scale in the polymer matrix. There is a great enhancement in tensile and dynamic mechanical properties of the nanocomposites. The tensile modulus and TS increase with the increase in organoclay content. Higher storage modulus over unfilled maleated PCL/soy protein blend was also observed. The storage modulus reflects the stiffness of the material. The enhancement of the storage modulus of the nanocomposite is due to the higher stiffness of the nanocomposite as compared with maleated PCL/soy protein blend.

3.4.4 Soy Protein Nanoblends

A kind of soy protein/polystyrene (PS) nanoblends with core-shell structures were successfully prepared by introducing nanosized PS into soy protein through emulsion polymerization [75]. The nanoblends showed core-shell structures, with the core being of PS and the shell of sodium dodecane sulfonate and soy protein polypeptides, as shown in Figure 3.7. The nanoblends had good thermoplastic properties and could be hot pressed into plastic sheets. Soy protein plastics possess good mechanical strength and water resistance by compositing PS nanoparticles into soy protein matrix. The water uptakes of the nanoblends ranged from 11–19%, which is much lower than that of pure SPI (32%) at 75% RH.

3.4.5 Biomimetic Nanocomposites

Through millions of years of evolution, nature has developed many interesting super hydrophobic surfaces, such as various plant leaves (a typical example being the lotus leaf) [76]. Surface roughness at a dual- or multi-length scale has shown to be the key in generating the surprising non-wetting behavior [77]. Inspired by the self-cleaning phenomenon found

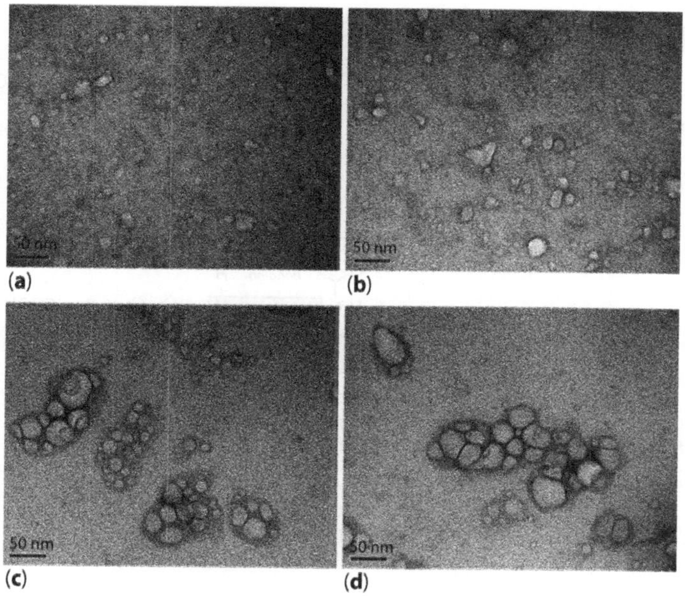

Figure 3.7 TEM micrographs of soy protein/polystyrene nanoblends with (a) 5%, (b) 10%, (c) 20%, and (d) 50% styrene contents.

in nature, the soy protein materials with high hydrophobic structures were prepared by immersing the arylated protein films (SB) in water for a long period of time (26 h) [78]. The author evidenced that water-induced microphase separation occurred on the surface and the interior of the water modified SB films (SBWM) after immersing SB films in distilled water for 26 h in the form of hydrophobic diphenylhydroxymethane (DPHM) shell. As shown in Figure 3.8, carboxylic groups of 2,2-diphenyl-2-hydroxyethanoic acid (DPHEAc) reacted with amino groups of soy protein to give arylated soy protein (I). Once these films were immersed in water, there was an evolution of CO_2, resulting in the formation of more stable compound DPHM (II). Furthermore, the hydrogen bonding between –OH groups of DPHM and $-NH_2$ groups of protein in the SBWM films created the hydrophobic DPHM shell on the surface of the protein chains as core. As shown in Figure 3.9, the SB films containing DPHEAc displayed as mooth structure on the surface, whereas the surface and the interior of the SBWM films after immersing SB in distilled water, exhibited nanospheres having a diameter of 30–40 nm, which were similar to the structure of some natural plant leaves such as lotus, rice, and taro [79]. It has been reported that the polymers having

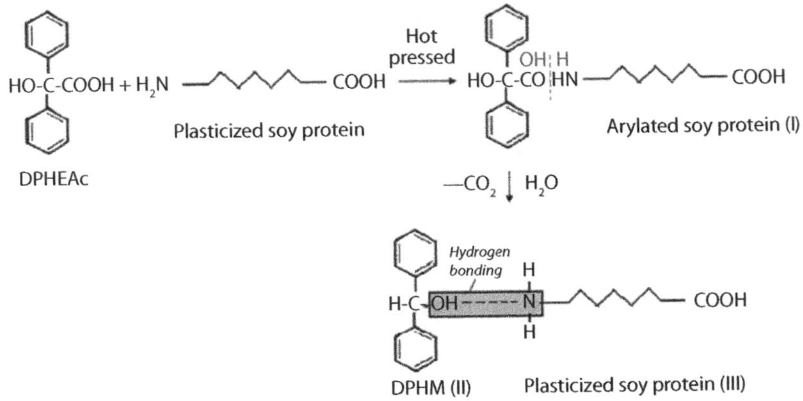

Figure 3.8 Reaction of arylated soy protein in water.

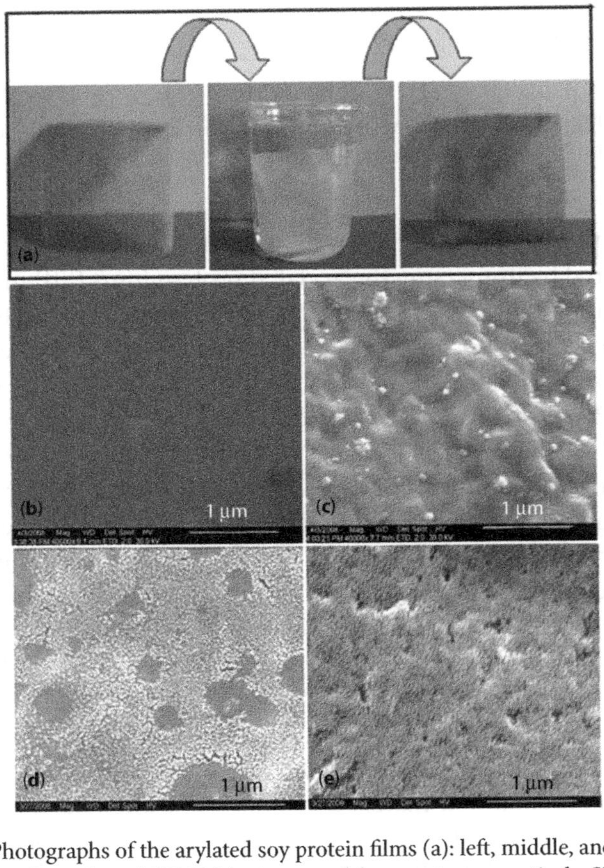

Figure 3.9 Photographs of the arylated soy protein films (a): left, middle, and right show SB, SB in water for 26 h, and SBWM just removed from water, respectively. SEM images of SB (b,c) and SBWM (d,e) films. Surface morphology (left) and cross-section (right).

natural lotus-like micro- and nanohierarchical structure possessed super hydrophobic surface [80]. Therefore, the lotus-like nanoscale spheres of the SBWM films also led to their hydrophobicity. SBWM films exhibited significantly higher contact angle than SB.

3.5 Conclusion

As a kind of agricultural processing by-product of the soybean oil industry, soy protein possesses good biodegradability, and finds wide applications as eco-friendly materials, edible films, and packaging films, etc. The incorporation of nanoparticle could improve the drawbacks of soy protein materials such as poor mechanical properties and high water sensitivity, and enlarge its application. The properties of the resulting nanocomposites are greatly dependent on the nature of nanofillers, the compatibility as well as the dispersion effect of the filler in the matrix. However, for the high water sensitivity of polyol plasticized soy protein, the water resistance properties of the nanocomposites may not be as good as synthetic polymers, such as polyethylene. Addition approaches, such as chemical modification of the matrix, blending soy protein with other hydrophobic polymers, would be involved with nanofilling. The incorporation of eco-friendly fillers would also be a trend for the studying and application of soy protein materials. Some fillers are not biodegradable and may be harm to human being, some green fillers from natural resources, such as polysaccharide based nanocrystals, would have a very bright future to fabricate totally biodegradable nanocomposites.

Acknowledgements

This work was supported by the National Natural Science Foundation of China (51203004), and Beijing Top Young Innovative Talents Program (2014000026833ZK13).

References

1. Thakur, V.K., Kessler, M.R., Free radical induced graft copolymerization of ethyl acrylate onto SOY for multifunctional materials. *Mater. Today Commun.*, 1(1–2), 34–41, 2014.
2. Wu, Q., Yoshino, T., Sakabe, H., Zhang, H., Isobe, S., Chemical modification of zein by bifunctional polycaprolactone (PCL). *Polymer*, 44(14), 3909–3919, 2003.

3. Petruccelli, S., Anon, M.C., Soy protein isolate components and their interactions. *J. Agric. Food Chem.*, 43(7), 1762–1767, 1995.
4. Khatib, K.A., Herald, T.J., Aramouni, F.M., MacRitchie, F., Schapaugh, W.T., Characterization and functional properties of soy β-conglycinin and glycinin of selected genotypes. *J. Food Sci.*, 67(8), 2923–2929, 2002.
5. Tian, H., Wu, W., Guo, G., Gaolun, B., Jia, Q., Xiang, A., Microstructure and properties of glycerol plasticized soy protein plastics containing castor oil. *J. Food Eng.*, 109(3), 496–500, 2012.
6. Guerrero, P., de la Caba, K., Thermal and mechanical properties of soy protein films processed at different pH by compression. *J. Food Eng.*, 100(2), 261–269, 2010.
7. Zhang, X., Burgar, I., Do, M. D., Lourbakos, E., Intermolecular interactions and phase structures of plasticized wheat proteins materials. *Biomacromolecules*, 6(3), 1661–1671, 2005.
8. Ly, Y.T.-P., Johnson, L.A., Jane, J., Soy protein as biopolymer. In: *Biopolymers from Renewable Resources*, D.L. Kaplan, Ed., Springer Berlin Heidelberg: Berlin, Heidelberg, pp 144–176, 1998.
9. (a) Chen, P., Zhang, L., Cao, F., Effects of moisture on glass transition and microstructure of glycerol-plasticized soy protein. *Macromolecular Bioscience*, 5(9), 872–880, 2005; (b) Chen, P., Zhang, L., New evidences of glass transitions and microstructures of soy protein plasticized with glycerol. *Macromol. Biosci.*, 5(3), 237–245, 2005.
10. (a) Liu, D., Zhang, L., Structure and properties of soy protein plastics plasticized with acetamide. *Macromol. Mater. Eng.*, 291(7), 820–828, 2006; (b) Liu, D., Tian, H., Zhang, L., Influence of different amides as plasticizer on the properties of soy protein plastics. *J. Appl. Polym. Sci.*, 106(1), 130–137, 2007; (c) Tian, H., Liu, D., Zhang, L., Structure and properties of soy protein films plasticized with hydroxyamine. *J. Appl. Polym. Sci.*, 111(3), 1549–1556, 2009.
11. Tian, H., Zhang, M., Liu, D., Liu, Q., Xiang, A., Chapter 14., Recent research on soy protein based blend materials, In: *Soy-based bioplasstic*, Vijay Kumar Thakur, Ed., Smithers Rapra Publisher, 2015.
12. Darder, M., Aranda, P., Ruiz-Hitzky, E., Bionanocomposites: a new concept of ecological, bioinspired, and functional hybrid materials. *Adv. Mater.*, 19(10), 1309–1319, 2007.
13. Liu, D., Tian, H., Chapter 4 Soy protein nanocomposites: emerging trends and applications. In: *Natural polymers*, Vol. 2: *Nanocomposites*, The Royal Society of Chemistry, pp. 91–112, 2012.
14. González, A., Alvarez Igarzabal, C.I., Nanocrystal-reinforced soy protein films and their application as active packaging. *Food Hydrocolloid.*, 43, 777–784, 2015.
15. Lu, Y., Weng, L., Zhang, L., Morphology and properties of soy protein isolate thermoplastics reinforced with chitin whiskers. *Biomacromolecules*, 5(3), 1046–1051, 2004.

16. Fernández-Espada, L., Bengoechea, C., Cordobés, F., Guerrero, A., Thermomechanical properties and water uptake capacity of soy protein-based bioplastics processed by injection molding. *J. Appl. Polym. Sci.,* 133(24), 43524, 2016.
17. Kumar, R., Choudhary, V., Mishra, S., Varma, I.K., Mattiason, B., Adhesives and plastics based on soy protein products. *Ind. Crops Prod.,* 16(3), 155–172, 2002.
18. Harper, J.M., In: *Extrusion cooking,* Mercier, C., Linko, P., Harper, J.M. Eds., American Association of Cereal Chemists: Minnesota, 1989.
19. Kumar, P., Sandeep, K.P., Alavi, S., Truong, V.D., Gorga, R.E., Preparation and characterization of bio-nanocomposite films based on soy protein isolate and montmorillonite using melt extrusion. *J. Food Eng.,* 100(3), 480–489, 2010.
20. Guo, G., Zhang, C., Du, Z., Zou, W., Tian, H., Xiang, A., Li, H., Structure and property of biodegradable soy protein isolate/PBAT blends. *Ind. Crops Prod.,* 74, 731–736, 2015.
21. Guo, G., Zhang, C., Du, Z., Zou, W., Xiang, A., Li, H., Processing and properties of phthalic anhydride modified soy protein/glycerol plasticized soy protein composite films. *J. Appl. Polym. Sci.,* 132(28), 42221, 2015.
22. Dash, S., Swain, S.K., Effect of nanoboron nitride on the physical and chemical properties of soy protein. *Compos. Sci. Technol.,* 84, 39–43, 2013.
23. Marney, D.C.O., Russell, L.J., Wu, D.Y., Nguyen, T., Cramm, D., Rigopoulos, N., Wright, N., Greaves, M., The suitability of halloysite nanotubes as a fire retardant for nylon 6. *Polym. Degrad. Stabil.,* 93(10), 1971–1978, 2008.
24. Nakamura, R., Netravali, A.N., Morgan, A.B., Nyden, M.R., Gilman, J.W., Effect of halloysite nanotubes on mechanical properties and flammability of soy protein based green composites. *Fire Mater.,* 37(1), 75–90, 2013.
25. Chattopadhyay, D.K., Webster, D.C., Thermal stability and flame retardancy of polyurethanes. *Prog. Polym. Sci.,* 34(10), 1068–1133, 2009.
26. Iman, M., Bania, K.K., Maji, T.K., Green jute-based cross-linked soy flour nanocomposites reinforced with cellulose whiskers and nanoclay. *Ind. Eng. Chem. Res.,* 52(21), 6969–6983, 2013.
27. Sasmal, A., Sahoo, D., Nanda, R., Nayak, P., Nayak, P.L., Mishra, J.K., Chang, Y.-W., Yoon, J.-Y., Biodegradable nanocomposites from maleated polycaprolactone/soy protein isolate blend with organoclay: Preparation, characterization, and properties. *Polym. Compos.,* 30(6), 708–714, 2009.
28. Chen, P., Tian, H., Zhang, L., Chen, Y., Wang, X., Du, Y., Structure and properties of soy protein/alumina hydrate nanocomposites fabricated via *in situ* synthesis. *J. Biobased Mater. Bioenergy,* 2(3), 248–257, 2008.
29. Robertson, G. L., *Food packaging: principles and practice, food science and technology.* 2nd edn., Taylor & Francis/CRC Press: Boca Raton, Fla, USA, 2006.
30. Schmid, M., Dallmann, K., Bugnicourt, E., Cordoni, D., Wild, F., Lazzeri, A., Noller, K., Properties of whey-protein-coated films and laminates as novel recyclable food packaging materials with excellent barrier properties. *Int. J. Polym. Sci.,* 2012, 7, 2012.

31. Rhim, J.-W., Park, H.-M., Ha, C.-S., Bio-nanocomposites for food packaging applications. *Prog. Polym. Sci.*, 38(10–11), 1629–1652, 2013.
32. Echeverría, I., Eisenberg, P., Mauri, A.N., Nanocomposites films based on soy proteins and montmorillonite processed by casting. *J. Membr. Sci.*, 449, 15–26, 2014.
33. Lee, J.E., Kim, K.M., Characteristics of soy protein isolate-montmorillonite composite films. *J. Appl. Polym. Sci.*, 118(4), 2257–2263, 2010.
34. Li, Y., Chen, H., Dong, Y., Li, K., Li, L., Li, J., Carbon nanoparticles/soy protein isolate bio-films with excellent mechanical and water barrier properties. *Ind. Crops Prod.*, 82, 133–140, 2016.
35. Tian, H., Zhang, L., Wu, Q., Wang, X., Chen, Y., Creation of hydrophobic materials fabricated from soy protein and natural rubber: surface, interface, and properties. *Macromol. Mater. Eng.*, 295(5), 451–459, 2010.
36. Sinha Ray, S.O.M., Polymer/layered silicate nanocomposites: a review from preparation to processing. *Prog. Polym. Sci.*, 28(11), 103, 2003.
37. Chaichua, B.P.S., *In situ* silica reinforcement of natural rubber by sol--gel process via rubber solution. *J. Sol-Gel Sci. Technol.*, 52(2), 9, 2009.
38. Tian, H., Processing and properties of soy protein/silica nanocomposites fabricated *in situ* synthesis. *J. Compos. Mater.*, 46(4), 9, 2012.
39. Ai, F., Zheng, H., Wei, M., Huang, J., Soy protein plastics reinforced and toughened by SiO_2 nanoparticles. *J. Appl. Polym. Sci.*, 105(3), 8, 2007.
40. Xia, C., Zhang, S., Shi, S.Q., Cai, L., Garcia, A.C., Rizvi, H., D'souza, N.A., Property enhancement of soy protein isolate-based films by introducing POSS. *Int. J. Biol. Macromol.*, 82, 6, 2016.
41. I. Mateos-Aparicio, A.R.C., Villanueva-Suarez, M.J., Zapata-Revilla, M.A., *Nutr. Hosp.*, 23, 305–312, 2008.
42. Khajavi, R.M.A., Electrospun biodegradable nanofibers scaffolds for bone tissue engineering. *J. Appl. Polym. Sci.*, 133(3), 19, 2016.
43. Rahman, M.M., Netravali, A.N., Tiimob, B.,Apalangya V., Rangari, V.K., Bioinspired "green" nanocomposite using hydroxyapatite synthesized from eggshell waste and soy protein. *J. Appl. Polym. Sci.*, 133(22), 11, 2016.
44. Rahman, M.A.B.V.K., Bioderived "green" composite from soy protein and eggshell nanopowder. *ACS Sustain. Chem. Eng.*, 2(10), 9, 2014.
45. Alexandre, M.P., Polymer-layered silicate nanocomposites: preparation, properties and uses of a new class of materials. *Mater. Sci. Eng. R*, 28(1), 63, 2000.
46. Kumar, P., Sandeep, K.P., Alavi, S., Truong, V., Gorga R.E., Effect of type and content of modified montmorillonite on the structure and properties of bio-nanocomposite films based on soy protein isolate and montmorillonite. *J. Food. Sci*, 75(5), 2010.
47. Wang, Z., Zhang, N., Wang, H., Sui, S., Sun, X., Ma, Z., The effects of ultrasonic/microwave assisted treatment on the properties of soy protein isolate/titanium dioxide films. *LWT - Food Sci. Technol.*, 57(2), 8, 2014.
48. Chen, P.L., Interaction and properties of highly exfoliated soy protein/montmorillonite nanocomposites. *Biomacromolecules*, 7(6), 7, 2006.

49. Jin, M., Plant protein-based nanocomposite materials: modification of layered nanoclay by surface coating and enhanced interactions by enzymatic and chemical cross-linking, *University of Tennessee, Knoxville*, 2012.
50. Jin, M., Zhong, Q., Transglutaminase cross-linking to enhance elastic properties of soy protein hydrogels with intercalated montmorillonite nanoclay. *J. Food Eng.*, 115(1), 8, 2013.
51. Jiahui Yu, Guojuan Cui, Ming Wei, Jin, H., Facile exfoliation of rectorite nanoplatelets in soy protein matrix and reinforced bionanocomposites thereof. *J. Appl. Polym. Sci.*, 104(5), 11, 2007.
52. Saito, R., Dresselhaus, G., Dresselhaus, M.S., *Physical properties of carbon nanotubes*, 1998.
53. Zheng, H.F.M.J.P.R., Thermoplastic soy protein nanocomposites reinforced by carbon nanotubes. *Macromol. Mater. Eng.*, 292(6), 9, 2007.
54. Arora, A.G.W., Review: nanocomposites in food packaging. *Journal of food science an official publication of the Institute of Food Technologists*, 75(1), 2010.
55. Chen, X.S.S., Titanium dioxide nanomaterials: synthesis, properties, modifications, and applications. *Chem. Rev.*, 107(7), 69, 2007.
56. Wang, S.-Y.B.-B.D.-Z.X.-Z.L., Preparation and characterization of TIO_2/SPI composite film. *Mater. Lett.*, 83, 2012.
57. Kim, J.S., Kuk, E., Yu, K.N., Kim, J.H., Park, S.J., Lee, H.J., *et al.*, *Nanomed.-Nanotechnol. Biol. Med.*, 3, 142–144, 2007.
58. Zhao, S.J.X.Z.X., Antimicrobial film by embedding *in situ* synthesized silver nanoparticles in soy protein isolate. *Mater. Lett.*, 95, 2013.
59. Tian, H.G., Processing and characterization of glycerol-plasticized soy protein plastics reinforced with citric acid-modified starch nanoparticles. *J. Polym. Environ.*, 19(3), 7, 2011.
60. Dufresne, K.G.N.a.A., *Biomacromolecules*, 4, 666, 2006.
61. Zheng, H.F.P.J.A., Structure and properties of starch nanocrystal-reinforced soy protein plastics. *Polym. Compos.*, 30(4), 7, 2009.
62. Wang, Y.X., Cao, X.D., Zhang, L.N., Effects of cellulose whiskers on propertiesof soy protein thermoplastics. *Macromol. Biosci.*, 6, 524–531, 2006.
63. Zhang, S.C.Y.Y.J.S.Q.C.L., Soy protein isolate-based films reinforced by surface modified cellulose nanocrystal. *Ind. Crops Prod.*, 80, 7, 2016.
64. Nakagaito, A.H., Novel high-strength biocomposites based on microfibrillated cellulose having nano-order-unit web-like network structure. *Appl. Phys. A-Mater. Sci. Process.*, 80(1), 5, 2005.
65. Huang, X.N.A., Biodegradable green composites made using bamboo micro/nano-fibrils and chemically modified soy protein resin. *Compos. Sci. Technol.*, 69(7), 7, 2009.
66. Arboleda, J.C.H.M.L.A.L.J.K.O.J., Soy protein–nanocellulose composite aerogels. *Cellulose*, 20(5), 2013.
67. Chen, G.H., Electrospun cellulose nanofiber reinforced soybean protein isolate composite film. *J. Appl. Polym. Sci.*, 110(2), 2008.

68. Rusa, C., Wei, M., Bullions, T.A., Rusa, M., Gomez, M.A., Porbeni, F.E., Wang, X., Shin, I.D., Balik, C.M., White, J.L., Tonelli, A.E., *Cryst. Growth Design*, 6, 1413, 2004.
69. Huang, J.Z.M.Y.P.R., Soy protein-based nanocomposites reinforced by supramolecular nanoplatelets assembled from pluronic polymers/β-cyclodextrin pseudopolyrotaxanes. *J. Appl. Polym. Sci.*, 107(1), 2008.
70. Hong, S.I., Choi, W.Y., Cho, S.Y., Jung, S.H., Shin, B.Y., Park, H.J., Mechanical properties and biodegradability of poly-ε-caprolactone/soy protein isolate blends compatibilized by coconut oil. *Polym. Degrad. Stabil.*, 94(10), 6, 2009.
71. Liu, B.L.H.L.J.W., Different effects of water and glycerol on morphology and properties of poly(lactic acid)/soy protein concentrate blends. *Macromol. Mater. Eng.*, 295(2), 7, 2010.
72. Guo, G.C.Z.W.H., Structure and properties of poly (vinyl alcohol)/soy protein isolate blend film fabricated through melt processing. *J. Polym. Environ.*, 23(2), 7, 2015.
73. Tian, H.L.Q.X.Y., Creation of hydrophobic materials fabricated from soy protein and natural rubber: surface, interface, and properties. *Macromol. Mater. Eng.*, 295(5), 9, 2010.
74. Sasmal, A.D.R.P.P.J.Y.J.Y., Biodegradable nanocomposites from maleated polycaprolactone/soy protein isolate blend with organoclay: preparation, characterization, and properties. *Polym. Compos.*, 30(6), 7, 2009,.
75. Liu, D.H.J.P.R., Core-shell nanoblends from soy protein/polystyrene by emulsion polymerization. *Macromol. Mater. Eng.*, 293(8), 8, 2008.
76. Blossey, R., Self-cleaning surfaces - virtual realities. *Nat. Mater.*, 2(5), 6, 2003.
77. Leng, B., Shao, Z., With, G. Ming, W., Superoleophobic cotton textiles. *Langmuir*, 25(4), 5, 2009.
78. Kumar, R., Zhang, L., Water-induced hydrophobicity of soy protein materials containing 2,2-diphenyl-2-hydroxyethanoic acid. *Biomacromolecules*, 9(9), 8, 2008.
79. Guo, Z. Liu, Biomimic from the superhydrophobic plant leaves in nature: Binary structure and unitary structure. *Plant Sci.*, 172(6), 10, 2007.
80. Xie, Q., Fan, G., Zhao, N., Guo, X., Xu, J., Dong, J., Zhang, L., Zhang, Y., Han, C.C., *Materials*, 1830–1833, 2004.

4

Applications of Soy Protein-Based Blends, Composites, and Nanocomposites

Ruann Janser Soares de Castro[1*], André Ohara[2], Paula Kiyomi Okuro [3], Camila Utsunomia[4], Joelise de Alencar Figueira Angelotti[5], Fabíola Aliaga de Lima[1,6] and Hélia Harumi Sato[1]

[1]*Department of Food Science, University of Campinas, Campinas, Brazil*
[2]*Division of Applied Bioscience, Hokkaido University, Sapporo, Japan*
[3]*Department of Food Engineering, University of Campinas, Campinas, Brazil*
[4]*Division of Biotechnology and Macromolecular Chemistry, Hokkaido University, Sapporo, Japan*
[5]*Institute of Chemistry, Federal University of Alfenas, MG, Brazil*
[6]*Chemical Engineering Department of the Escola Politécnica, University of São Paulo,SP, Brazil*

Abstract

Soybeans are the dominant oilseed in global markets been one of the most consumed grains in the world. This dominance comes from several aspects: favorable agronomic characteristics, reasonable profit to producers, high-quality protein meal for animal feed, high-quality edible oil products, and plentiful availability at competitive price. Despite these traditional features, special applications of soy protein isolate (SPI) for development of biomaterials, composites, nanocomposites, and blends have been reported as potential use of this grain in several areas, such as biomedical, optoelectronic, optical coating, and packaging. In regards to SPI particulars, the usage in the medical field presents several applications as drug delivery systems, tissue engineering, replacement implants, and wound dressing. Those SPI particulars can also be used in whiskers fibers to produce liquid crystals and electrolytes for lithium batteries. In addition thin films can be developed with SPI particulars and be applied in optical coatings (CD and DVD, projection TV, TFT LCD display) as well as food packing. Another feature is that SPI can be molded into blends with other materials forming films with distinct applications: bioplastics, wood adhesives, edible films, and biomedical functional membranes.

**Corresponding author*: ruannjanser@hotmail.com

Visakh P. M and Olga Nazarenko (eds.) Soy Protein-Based Blends, Composites and Nanocomposites, (67–102) 2017 © Scrivener Publishing LLC

Recently SPI has been applied in nanocomposites that can be used in aerospace and automotive structures, packing films and coatings for pharmaceuticals and food, and electrospun fibers for biomedical engineering. In summary SPI presents enormous potential for development of biomaterials with several applications, also the use of soy protein in large scale would be very important for the environment issue since it is an abundant, cheap, versatile, and biodegradable material.

Keywords: Soy protein isolate, edible films, nanocomposites, biomaterials

4.1 Introduction

Soybeans have been cultivated for more than 3000 years in China and other Asian countries, such as Japan and Korea. According to USDA [1], the world production of soybeans in 2013 has in leading the United States and Brazil (32% and 31%, respectively), followed by Argentina (19%) and only 4% of this production is in China.

The typical composition of the soybean is oil (18%), protein (38%), insoluble carbohydrate (dietary fiber) (15%), soluble carbohydrate (as sucrose, stachyose, raffinose, and others) (15%), and ash (14%) [2]. Soybean protein can to be classified in three types: (1) dehulled soybeans that are used to make full soy flour; (2) soy protein concentrates (SPCs) that have a dry weight protein content of at least 70% by definition; (3) soy protein isolates (SPIs) that have a dry weight protein content of at least 90% by definition [3].

Soybeans have been important raw material for food industry, because is one of the most consumed grains in the world. However, special applications of soy protein for development of biomaterials, composites, nanocomposites, and blends have been reported as potential use of this grain in several areas, such as biomedical, optoelectronic, optical coating, and packaging [4–14]. This is possible due to the ease of processability of soy protein, abundant resource, low cost, specific mechanical properties, and environmental friendliness [15–17].

The study of soy protein in these applications is interesting due to its physicochemical characteristics allowing materials to be thermoplastic, to form hydrogels, adhesive capacity, low thermal expansion coefficient, optical transparency, anisotropy, among other properties [18].

Processes involving the development of composites and nanocomposites have been highlighted as relevant applications of soy protein. Biodegradable composites from natural materials (such as soy protein) provide many environmental benefits because it reduces wastes and the need to synthesize petroleum-based polymers, increases energy efficiency, promoting the concept of sustainability [12, 15]. The soy protein nanocomposites can be used as adhesives, asphalt, resins, cleaning products, paper coatings, plastics, polyesters, and textile fibers that allow its use over a large area, such as packaging, medical, aerospace, and automotive fields [7, 19, 20].

In this context, the aim of this chapter was to review the literature on the applications of soy protein-based blends, composites, and nanocomposites and their potential of use.

4.2 Applications of Soy Protein Particulars

4.2.1 Biomedical Applications

Soy proteins possess several advantages in the medical field since its non-animal origin eliminates the risk of transmissible diseases which poses a danger in human and animal-derived products. It has been reported that soybean-based products promote tissue regeneration, such as new bone growth. Besides these materials can integrate into blood clots and stimulate collagen deposition and also stimulate cells to produce new tissue with no need for expensive growth factors having therefore, significant potential for wound-healing applications [21].

Numerous soy protein-based materials have been studied and can be classified in several groups of plastics, gels, films, and additives or coatings. However, most of these soy protein-based materials are modified chemically or blended with other synthetic/natural polymers to overcome their fundamental limitations, such as water sensitivity, poor processability, and low mechanical strength [22].

Regarding biomedical applications, soy protein-based blends, hydrogels, and membranes are the most studied and successfully applied. Several applications of soy protein-based blends, composites, and nanocomposites will be covered in the following subsections of this chapter.

4.2.1.1 Drug Delivery Systems

Controlled drug delivery systems have been widely studied with the aim of overcome the shortcomings of conventional drug formulations. Although significant progress has been made in the controlled drug delivery area, more advances are yet to be made for treating many clinical disorders, such as diabetes and rhythmic heart disorders [23]. Chen *et al.* [24] evaluated the potential of SPI films as controlled release systems for active compounds. Thus mechanical properties, dissolution, and compound release kinetics of SPI films were analyzed and demonstrated an excellent compound binding capacity, especially for hydrophobic molecules, and hence potential for use in controlled release systems based on matrix erosion. Vaz *et al.* [5] investigated the drug encapsulation into a soy matrix by melt extrusion and evaluated its suitability to be used as a drug delivery system. This new protein matrix showed a pH-sensitive behavior, which increases the possibility for application as controlled delivery devices to carry a range of bioactive agents. Vaz *et al.* [25] also studied a double-layer delivery device based on soy protein produced by using an innovative co-injection moulding technique and related the potential of this system to achieve a controlled drug delivery.

Soy protein gels of filamentous or particulate microstructure were also studied by Maltais *et al.* [26] as a nutraceutical delivery device for riboflavin. This hydrogel demonstrated itself as a useful transportation for bioactive molecules through the gastrointestinal tract.

4.2.1.2 Tissue Engineering

The current demands for transplant organs and tissues are certain overcoming the supply, and all projections indicate that this gap will continue to increase, in this scenario emerges the tissue engineering, which is an important emerging topic in biomedical engineering and has shown tremendous promise in creating biological alternatives for harvested tissues, implants, and prostheses [4].

The successful application of regenerative medicine relies on the availability of biodegradable biomaterials, which participate in the biological processes that promote the formation of new tissue. Therefore, the repair of soft and hard tissues has been studied using both synthetic and natural polymeric biomaterials (sometimes blends composed by both), trying to reach the ability to exert a biospecific activity on the tissue cells, that is, cell differentiation and stimulation of a new extracellular matrix secretion [27].

4.2.1.2.1 Replacement Implants

A promising study regarding the usage of soy protein particulars was carried out by Vaz *et al.* [28] who reported casein and soybean protein-based thermoplastic proposed for biomedical usage, which presented a suitable range of mechanical and degradation properties, as well as a bioactive character (especially when reinforced with bone-like ceramics) which is desirable for biomedical applications, such as in the orthopedic field. This result represents an important advance in the replacement implants area; according to Swetha *et al.* [29] natural polymers often possess highly organized structures which can guide cells to grow at various stages of development stimulating an immune response at the same time thus, several natural polymers have been reported for their applications in bone tissue engineering.

4.2.1.2.2 Wound-Dressing Applications

Regarding wound-dressing application, Meikle *et al.* [27] studied soybean-based hydrogels with different physicochemical properties and bioactivity, which were obtained by sequential or simultaneous procedure of soy flour defatting process and material extraction. Interesting results were evidenced; the biomaterials obtained can be used as either bioadhesives or injectable formulations in regenerative medicine as they stimulated the synthesis of collagen by fibroblasts and the formation of mineralized bone noduli by osteoblasts. Another promising study was performed by Peles and Zilberman [21] who combined the wound dressing and drug delivery properties of a soy protein-based film characterizing its mechanical and physical properties. The results demonstrated good mechanical properties with the desired physical properties and controlled release of the antibiotic drug gentamicin and can, therefore, be potentially useful as burn and ulcer dressings.

The literature relates several studies regarding the application of soy protein in the biomedical field, which can be found on the Table 4.1.

4.2.2 Optoelectronic Applications

SPI the major component of soybean has been used to prepare biodegradable materials, such as adhesives, plastics, and various binders in recent years. Although the SPI plastics usually possess good biodegradability, their application is limited by poor flexibility and water resistance. The brittleness can be controlled easily by using plasticizers, but their use unavoidably leads to a significant decrease in the tensile strength. To obtain flexible SPI plastics with a high tensile strength, several kinds of filler such as cellulose or chitin whisker have been studied [39].

Whiskers are fibers which have been grown under controlled conditions that lead to the formation of high-purity single crystals. They constitute a generic class of materials having mechanical strengths equivalent to the binding forces of adjacent atoms. The resultant highly ordered structure produces not only unusually high strengths but also significant changes in electrical, optical, magnetic, ferromagnetic, dielectric, conductive, and even superconductive properties. The tensile strength properties of whiskers are far above those of the current high-volume content reinforcements and allow the processing of the highest attainable composite strengths [40].

Compared with inorganic whiskers, whiskers from renewable resources have advantages such as renewability, low cost, easy availability, good biocompatibility, and easy modification chemically and mechanically. For example, cellulose nanoparticles can be extracted from lignocellulosic

Table 4.1 Applications of soy protein-based blends and composites in the biomedical field.

Structure	Composites	Technique	Application	Reference
Solid matrix	SPI	Extrusion and injection moulding	Controlled drug release	[5]
Film	SPI	Solvent casting method	Wound dressing and drug delivery	[21]
Film	SPI	Stirred in as a plasticizer with glycerol	Controlled drug release	[24]
Double-layer delivery device	SPI	Co-injection moulding	Controlled drug release	[25]
Hydrogel	SPI	Cold gelation	Controlled drug release	[26]
Hydrogels	SPI	Soy flour defatting process	Regenerative medicine	[27]
Thermoplastic	Casein/SPI	Co-rotating twin-screw extruder	Orthopedic field	[28]
Hydrogel	Soy protein/poly acrylic acid	Free radical copolymerization	Controlled drug release	[30]
Blend	SPI/poly ethylene oxide	Electrospum fibers	Controlled drug release	[31]
Porous structure	Chitosan/SPI	Freeze-drying	Scaffold for tissue engineering of cartilage	[32]
Sponges	Cellulose/SPI	Freeze-drying	Tissue scaffolds, implantation devices, and drug delivery device	[33]
Hydrogels	SPI	Bioartificial gel technologies	Wound dressing	[34]
Membrane	Cellulose/SPI	Casting and coagulation	Tissue engineering	[35]
Membrane	Chitosan/SPI	Solvent casting	Membrane-based tests	[36]
Membrane	Casein/SPI	Stirred in as a film with glycerol	Drug delivery and wound repair systems	[37]
Membrane	Cellulose/SPI	Blending in NaOH and thiourea aqueous solution	Vero cells culture	[38]

fibers and are available all around the World; Chitin from shellfish, insects, and microorganisms is the second most abundant structural biopolymer. Cellulose and chitin whiskers have been used as a new kind of nanofiller as a reinforcing phase in both synthetic polymeric matrixes and natural ones [41].

In the nanocomposite field, up to now, cellulose whiskers were only used as geometrically and structurally well-defined model cellulosic filler and no practical industrial application was envisaged. This is mainly ascribed to the duration of the preparation technique. However, two important applications may be cited. The first one concerns the liquid crystals for optical applications like in security paper. The second is related to the use of these whiskers as mechanical reinforcing agents for low-thickness polymer electrolytes for lithium batteries application [40].

4.2.2.1 Liquid Crystals

Liquid crystals represent a state intermediate between ordinary liquids and three-dimensional solids; the investigation of their physical properties is very complex and makes use of many different tools and techniques. Liquid crystals are also model materials for the organic chemistry in order to investigate the connection between chemical structure and physical properties, and they provide insight into certain phenomena of biological systems [42].

Cellulose, one of the most versatile and widely found biopolymers in nature, has been used by humans for millennia as a building material, an energy source, a component of clothing and for storing and sharing knowledge and culture. One strong trend, on an international scale, is to focus on the isolation of fibrils and whiskers of cellulose with diameters in the nanometer range and to utilize their enhanced properties to develop novel cellulose-based materials with diverse advanced functionalities [43].

The application of cellulose nanoparticles to develop liquid crystals has been widely studied by Goldbout *et al.* [44], Werbowyj and Gray [45], Werbowyj and Gray [46], Gray [47], Gilbert and Patton [48], Bhadani and Gray [49], and Lagerwall *et al.* [43], however the combination process of nanofilling a matrix, for example, filling a SPI matrix with cellulose whiskers, has not been regularly studied. This approach offers a promising technology regarding the liquid crystals construction since it aggregates attractive properties of both compounds such as biocompatibility, a high elastic modulus (similar to steel), a low thermal expansion coefficient, optical transparency and anisotropy, negative diamagnetic anisotropy, flexible surface chemistry and low price.

As an example, in studies performed by Lu *et al.* [41] and Anglès and Dufresne [50] nanocomposites materials were successfully developed using a colloidal suspension of chitin whiskers as a filler to reinforce (SPI) plastics. Also another material was developed; a glycerol plasticized starch was used as the matrix and a colloidal suspension of cellulose whiskers as the reinforcing phase. It was demonstrated that incorporating whiskers into matrix leads to an improvement in water resistance which makes them interesting for industrial application due to the high performance of the resulting composites.

4.2.2.2 Electrolytes for Lithium Batteries

Lithium batteries are characterized by high specific energy, high efficiency, and long life. These unique properties have made lithium batteries the power sources of choice for the consumer electronics market with a production of the order of billions of units per year. These batteries are also expected to find a prominent role as ideal electrochemical storage systems in renewable energy plants, as well as power systems for sustainable vehicles, such as hybrid and electric vehicles [51].

Polyoxyethylene-based polymers are the most frequently used materials for polymer electrolyte applications in lithium batteries due to their cationic solvatation ability. However, it was demonstrated that tunicin (a substance present in the tunicates, which is identical to cellulose) whiskers strongly increased mechanical properties of polyoxyethylene-based polymers. The processing of a composite polymer electrolyte from an aqueous suspension of cellulose whiskers is not easy to consider since water can react with the negative electrode and reduce the battery cycle life. Therefore, the main obstacle for the use of cellulose whiskers as a reinforcing phase in a wide variety of polymers is the limited stability of whiskers in aqueous suspension [52].

Several recent studies relate the application of cellulose in electrolytes for lithium batteries [53–56], however a combination process as the reinforcement or incorporation of SPI with cellulose whiskers aiming the application in electrolytes for lithium batteries has not been regularly studied. Equally comparing the liquid crystals sub item this approach offers a promising technology since it unifies attractive properties of both compounds. In study performed by Wang *et al.* [39] in which they incorporate cellulose whiskers to SPI aiming to improve mechanical properties, the authors obtained composites which showed greater water-resistance and thermal stability. The improvement in the properties of the SPI/cellulose whisker composites may be ascribed to cross linking networks caused by intermolecular hydrogen bonds between the cellulose whiskers and the SPI matrix.

4.2.3 Optical Coatings

A thin film is a system which consists of one or more homogeneous or heterogeneous, absorbent or non-absorbent layers deposited onto a suitable substrate. Thus, optical thin films are widely used and increasing their importance in nearly all technologies of optics and photonics. They not only improve the optical performance of optical devices but also are vital parts in optical operations. There are a several applications of thin films coatings in optics and photonics: cell phones, lattices, digital cameras, binocular telescopes, CD and DVD, projection TV, TFT LCD display, fax machines, eyeglasses, sunglasses, telecom, LED, organic LED, solar cells, smart windows, green buildings, security, satellites, microscopes, etc., covering the spectral ranges from IR to EUV [57, 58].

Regarding the usage of soy-based particulars, several studies relate the usage of epoxidized soybean oil and its application in the optical and photonics field [9, 11, 59–62]. However this chapter is focused on soy proteins and its functionalities, therefore those works will not be discussed.

The application of soy protein particular in the optical coating has not been regularly studied; nevertheless this protein presents interesting characteristics for this area as high elastic modulus, a low thermal expansion coefficient, optical transparency and anisotropy. As an example, Ji *et al.* [18] elaborated a poly (ethylene oxide)/soy protein film and obtained a fully amorphous structure with ultra-elasticity. Other promising results were showed; the ionic conductivity was dramatically enhanced compared with that of pure poly(ethylene oxide)-based film. This amorphous thin films are easy to fabricate, and both processing and material usages were environmental friendly, offering great potential for applications in coating foldable/flexible electronics.

Another interesting and recent application is related to carbon nanotubes which offer a unique combination of electrical, mechanical, thermal, and optical properties that make them highly promising materials for numerous applications. In study performed by Ji *et al.* [63] carbon nanotubes were treated with SPI for improve the dispersion. The authors demonstrated interesting results; soy proteins were beneficial to increase the interaction between carbon nanotubes and the polymer matrix. Also good properties, such as mechanical properties, could be expected. Additionally, this finding links one of the central molecules in biology, SPI, to the most important 1D nanofiller, carbon nanotubes open up a new field for the application of rich soy products.

4.2.4 Packaging Applications

Biopolymers are associated with good film forming properties which makes them proper to packaging applications [64]. Nevertheless natural polymers showed poor mechanical and "barrier" properties besides low thermal stability. Different strategies are adopted to overcome these limitations. Some of them are based on blends with synthetic and biodegradable polymers such as polysaccharide-protein mixtures, new composite materials, application of plasticizers, chemical modifications like cross-linking, nanocrystal-reinforced systems and others [15, 22, 65–67].

Functional properties of the soy protein films are mainly determined by their microstructure (protein-protein interactions such as disulfide covalent bonds, hydrogen-bonding, electrostatic attractions, and hydrophobic bonding) that is related to film-forming process experimental parameters used in films preparation and protein structure according with amino acid composition and molecular weight, primary structure, protein configuration, protein charge distribution, extension of intra- and intermolecular bonding (quaternary structure), and the environment [68–69].

The protein concentration in film-forming solutions is an important variable for the film characteristics. Soy protein can be classified based on protein content, where SPI shows the higher concentration of protein, 90%, followed by the soy protein concentrate (SPC) that contains 65 up to 72% of protein and the defatted soy flour (DSF) with 50% upto 59% [70]. The mechanical properties, tensile strength (TS), and percentage elongation at break (E), increased as protein content in the film forming solutions increasing (from DSF to SPI), it can be explained by the presence of insoluble components in DSF and SPC [70]. SPI is a biopolymer that has potential applications in packaging (films or coatings), because offers interesting film-forming properties, good barrier properties to oxygen, aromas and lipids when in low to intermediate moisture conditions, besides it is a low cost raw material [71].

Another important variable in soy protein film formation is pH of the system. SPI films can be formed at both alkaline and acidic medium, but that one formed at alkaline conditions (pH 8 up to 11) showed greater values for tensile strength and percentage elongation at break and worst values for the water vapor permeability [71]. Guerrero and de la Caba [68] studied the pH effect on physicochemical properties; they investigated film properties pH dependence using Fourier transform infrared (FTIR) spectroscopy, differential scanning calorimetry (DSC), and thermogravimetric analysis (TGA). Films were produced by freeze-drying followed by compression at different pHs: acid pH (pH = 1.4), at the isoelectric pH (pH = 4.6), at the unmodified pH (pH = 7.5), and at a basic pH (pH = 10.0), using glycerol as plasticizer. The results shown that both tensile strength and elongation at break are higher at basic pHs, it can be explained by the protein structure, the unfolding configuration (allowed the exposition of the polar groups) seems to be appropriated in order to favor the protein-glycerol interaction, and thus contributing with the better mechanical properties observed.

One of the formation processes of soy protein films has been described involving two-step: heat denaturation of the proteins followed by surface dehydration, a network is formed when the unfolded proteins are able to interact through disulfide bonds and hydrophobic interactions [72–74]. Based on those film-forming fundamental, Kokoszka *et al.* [74] proposed the understanding of the effect of both soy protein and glycerol contents on physico-chemical characteristics of SPI-films. The authors observed a synergistic effect on the water vapor permeability of glycerol and soy protein content.

Soy protein material is able to produce softer, more transparent and flexible films when compared to those derived from other sources [75]. However due to hydrophilic nature of soy protein some disadvantages can be pointed: poor properties of moisture barrier and relative fragility when in wet state [15]. In the other side some chemical interventions can be applied in sense of change some physicochemical characteristics of the material [76].

In fact, explore combinations between soy protein and other biopolymers is an attractive film-forming material owing to its linkages that can improve mechanical properties of the material. Piazza *et al.* [77] studied the surface aggregating soy protein and high methoxyl pectin (HMP) solutions in order to produce biodegradable films for food packaging applications aiming better protection performance than pure soy protein. The film was developed at the air-water interface. The interfacial layer formed by pure SPI showed small regime of linear viscoelasticity response, and there is no correlation between the accumulation of SPI with the higher surface coverage, owing to the time required by the protein to migrate at the interface to unfold. When the HMP was added to forming a composite system SPI-HMP at air-water interface there was an increase in the elastic interfacial modulus, due to control of the protein complexation at the interfacial area, contributing with the surface layer stability. The composite system stability was attributed to the formation protein–polysaccharide complex at the air-water interface.

The bonds and interactions involved between protein chains characterize protein-based films formed as fragile and brittle material. Regarding that soy protein is formed by polar and nonpolar side chains and then allowing hydrogen bonding, dipole-dipole, charge-charge, and hydrophobic interactions. The polar interactions between side chains of soy protein molecules determined the poor rotation and molecular mobility increasing the stiffness, yield point, and tensile strength that characterizes brittle soy protein plastics. The use of plasticizers can overcome this characteristic, reducing interactions between protein chains and increasing the flexibility, extensibility, and processability features [78]. Desirable plasticizers characteristics such as small size, high polarity, presence of more than one polar group are related with the better performance in the plasticizer-protein interaction, where biopolymeric films are plasticized by hydroxyl compounds [68, 79]. There are some theories to explain the mechanisms involved in the plasticizer effect, among them are the lubricity, gel and free volume theory, all of them are based on the fundamental concept where the plasticizer can interpose itself between the polymer chains in this way decreasing the forces that holding these chains together [80]. Thus plasticizer agents are able to break the intermolecular linkage of the native protein structure, leading to a new orientation and restructuration of the chains, as well as permits the formation of new intermolecular linkage that can stabilize the reorganized three-dimensional network formed [68].

Commonly the plasticizer agent works reducing cohesion within protein film, improving chain mobility replacing protein-protein interactions by protein-plasticizer interactions, extending and softening the film structure due to association in molecular polymeric chains this way prevent cracking [81, 82]. Although a problem must be considered, in fact increasing mobility of the chains causes an increase in the diffusion coefficient of water vapor and oxygen, leading to lower the mechanical strength of the material, so it is important optimize the content of plasticize agent to obtain desirable mechanical strength film and reduce undesirable effect on its properties [69, 83–84]. The commons plasticizers used include polyols, such as sorbitol, glycerol, and glucose [85–88].

Guerrero *et al.* [89] blended SPI with gelatin and glycerol as plasticizer, but in this case the films were prepared by compression molding. As result flexible and transparent films were obtained. The addition of gelatin (15%) improved the tensile strength and the elongation at break, decreased the hydrophilicity and maintained the UV barrier properties when compared with the ones without gelatin. Another pointed strategy for enhanced the behavior of soy protein packaging systems is the modification of SPI with cross-linking agents. Cross-linking agents can work improving the mechanical and water barrier properties of soy protein films reducing its solubility, ability to swell and gas/water vapor permeability. A protein network as result of a greater degree of cross-linking characterize better microstructures density reduces the solubility properties and enhanced mechanical strength and barrier qualities [69, 90–91]. The degree of cross-linking and the hydrophilic-hydrophobic film profile were determined by the protein-protein interactions involved in film formation [68].

González *et al.* [15] produced cross-linked soy protein as material for biodegradable films using a nontoxic cross-linker, genipin. The tests performed stated the improvement in the film properties such as decrease in solubility, swelling and water vapor permeability, and mechanical properties (tensile strength and elongation at break). The SPI cross-linked with genipin was considered promising natural biodegradable materials for use in food packaging. As mentioned numerous strategies have been studied in order to improve water barrier ability of biofilms, including the addition of polysaccharides, proteins, or lipids. Inspired for this possibility Guerrero *et al.* [92] produced protein-based films prepared with acids and oils, lactic acid, epoxydized soybean oil and olive oil, by compression at pH 10.

The processing method used to produce the soy protein-based films plays a role in determining the physical and mechanical properties. Many processing methods are described to produce packaging materials with soy protein based in thermo-mechanical processing or dissolution-solvent evaporation, intensive mixing followed by compression molding and solution-casting [69]. Many inspirations about use soy protein to produce biodegradable and/or edible packaging systems have been studied with diverse components and forming process as showed in Table 4.2.

4.3 Applications of Soy Protein-Based Blends

Soy proteins are rich in polar groups, they can be molded into blends with other materials and yield resins that possess performance characteristics comparable to those of commercial thermoplastics [10, 95–96]. Blending is cost-effective and one of the most useful methodologies to improve properties of SPI-based materials [13, 97–98]. It is considered an important approach because for industrial applications SPI films do not show good mechanical and barrier properties without a secondary component [99]. However, some limitations need to be overcome for obtaining blends with desirable properties. One of the issues is obtaining acceptable adhesion with the other components of the blend, besides thermal and environmental stability of the materials also may restrict the field of application [95].

Table 4.2 Production of packaging based on soy protein raw material.

Packaging materials	Components	Properties	Reference
Edible oxygen barrier bilayer film pouches	SPI and corn zein	Higher oxygen barrier property	[73]
Active biodegradable bilayer films	SPI, lignin and formaldehyde	High water resistance and malleability	[76]
Biodegradable films intended for food packaging applications	Surface aggregating soy protein-high methoxylpectins solutions	For composite system better surface layer stability than pure SPI	[77]
Biodegradable films	SPI, bovine gelatin type A and glycerol	Flexible and transparent	[89]
Biodegradable bilayer films	Isolated soy protein (SPI) and poly (lactic acid) (PLA)	High transparency and strong adhesion between layers	[93]
Biodegradable hot-melt adhesive	Poly-ε-caprolactone (PCL), (SPI) and plasticizers: coconut oil and PEG400	The thermal and mechanical properties of the blends changed with increasing SPI concentration.	[94]

Thus, it will be shown in the next sub items that many efforts have been taken for obtaining soy protein blends with enhanced properties for their use in many fields of application.

4.3.1 Bioplastics

Biopolymers derived from natural resources are considered as potential substitutes for existing petroleum-based synthetic polymers due to their low cost, easy availability from renewable resources, and biodegradability [10]. Among the biopolymers derived from natural resources, renewable resources from farm products, such as starches and proteins, are more naturally abundant and economically feasible to use than many other biopolymers such as poly(lactide) or poly(hydroxybutyrate) [95, 100, 101]. Nevertheless, the utilization of proteins directly as engineering materials faces several challenges, since the mechanical properties of cereal proteins are not as robust as those of traditional thermoplastics [95]. Therefore, many efforts, such as blending with biodegradable polyesters [95] and with gelling agents [102], have been made to enhance the mechanical properties and decrease the water sensitivity of the soy protein-based plastics. Li *et al.* [10] have chosen poly(butylene succinate) (PBS) to blend with SPI, a thermoplastic polyester with desirable mechanical characteristics. In this study, as a strategy to improve the compatibility between SPI and PBS, the polyester was pretreated with different amounts of urethane and isocyanate groups, providing much finer phase structures because of good dispersion of polyester in protein. It was observed a lower glass transition temperature of protein in the blends than that of the pure SPI. The water resistance of the soy-protein-based plastics was also significantly improved. In a different way to explore the advantages of blending, Liu *et al.* [103] blended poly(lactic acid) (PLA) with SPC in an attempt to lower PLA overall cost and improve some of its properties. In this work, were investigated the effects of two different processing aids on the processing properties of the PLA/soy protein concentrate (SPC) blends. Both processing aids decreased tensile strength and modulus of the blends and increased break strain and impact strength.

4.3.2 Wood Adhesives

Currently formaldehyde-based wood adhesives predominate in the market and most of them use petrochemicals as raw materials [104]. Due to the tightened environmental regulations on the emission of volatile organic compounds, that causes both environmental and health problems, and also to decrease our heavy dependence on petrochemicals, many efforts have been made to develop formaldehyde-free wood adhesives from renewable natural resources [104, 105]. Soy protein has many unique properties such as low cost, ease of handling, low press temperatures, and the ability to bind wood with relatively high moisture content, representing a very practical and inexpensive material for wood adhesives [104]. However, soy protein adhesives low adhesion strength and water resistance greatly limit their practical applications [106], therefore by blending soy protein adhesives with other protein [107] or synthetic adhesives [104, 108–110] it is possible to generate products with enhanced performance [110, 111].

In order to improve the adhesion strength of soy protein adhesives on wood veneer, Xiao *et al.* [106] developed a blend of sorghum lignin or extruded sorghum lignin with soy protein adhesives (SPAs) based on SPI or modified soy protein (MSP). The main source of lignin is the liquid waste from paper pulp production, and this material is the unique nonpetroleum resource that can provide renewable aromatic compounds [112]. It is believed that the generation of strong bonding effects depends on coupling reactions of side chains, link-bonds, and functional groups, therefore this group verified that shear strength and water resistance of soy protein adhesives can be improved by adding lignin. In a patent of Thames *et al.* [111] was developed a water resistant soy protein-based adhesive blend with polyol plasticizer, preferred glycerol, and with a vegetable oil derivative, aleinized methyl ester of tung oil. This adhesive can be useful in the manufacture of particleboard and other composites.

4.3.3 Edible Films

Edible films, as explained in previous sub item of the present chapter, are flexible materials used in food coating and packaging [8]. They can control moisture, gases, and lipid migration extending the shelf life of foods, and being supporters of additives and nutrients [113, 114]. Furthermore, they can be an alternative to reduce the cost and also the amount of traditional packaging; a promising solution to environmental impacts of synthetic polymer packaging [115]. Since SPI-based films have excellent aroma, oxygen and lipid barrier properties, they can be used as edible packaging materials [116], mainly for applying to sensitive-to-oxygen products or for preserving flavors [117]. Nevertheless, SPI films high water vapor permeability is the most important obstacle to its use in food coatings [86], drawback which can be exceeded by blending methodology and thus many works have been developed in this area by blending, that is, with cassava starch [114], poly(vinyl alcohol) (PVA) [118] and lauric, palmitic, stearic, and oleic acid [119]. It should be emphasized that for edible films, materials used for blending with soy protein must be safe to eat or generally recognized as safe (GRAS), compatible with SPI and have the ability to form films [97].

Carboxymethyl cellulose (CMC) is one of the natural water-soluble cellulose derivatives, used as a highly effective additive to improve product and processing properties. Due to its nontoxicity, biocompatibility, biodegradability, hydrophilicity, and good film-forming ability, CMC has been used in a number of edible film formulations [120]. For these reasons Su *et al.* [97] created edible films based on CMC and SPI compatibilized by glycerol. CMC and SPI formed one phase blends and was observed that increasing the CMC content improved the mechanical properties and reduced the water sensitivity of blend films.

It is known that lipids or more hydrophobic proteins can be used in composite films to improve the film's ability to control water transport. In addition some studies revealed that films from sodium and calcium caseinates showed better water barrier properties although their films are less flexible and stretchable [121]. Therefore, to improve SPI film properties Monodero *et al.* [117] created SPI-based films containing sodium caseinate (Na-Cas) or calcium caseinate (Ca-Cas), with and without lipids (oleic acid (OA) and beeswax (BW). Caseinate incorporation to SPI-based films provoked an increase of elastic modulus and tensile strength at break and also enhanced the transparency of SPI-based films and color softening. Both caseinates contributed to increase the water vapor barrier properties of soy protein-based films.

4.3.4 Clinical Medicine and Pharmacy Field

Synthetic, natural, or biological materials have been used along the years as key elements in diverse biomedical applications, such as controlling/modulating the healing mechanisms and the outcomes of wound repair upon dressing [122]. However, synthetic polymers such as polyurethane, polyethylene, polylactides, polyglycolides, and polyacrylonitrile have disadvantages in such applications due to poor biocompatibility and release of acidic degradation products [36]. In this context, for the nutritional and health benefits of soy protein it has been proposed as a non-animal origin protein substitute for several biomedical applications whereby tissue-engineering scaffolds, wound-dressing material, and drug delivery system [34, 123]. This material has the advantages of storage stability, economically competitive, similarity to tissue constituents and reduced susceptibility to thermal degradation [124]. Thereby, through blending it is possible explore the bioactivities of soy proteins and at same time enhance its water resistance, flexibility and others mechanical properties. With the aim to develop a biodegradable material with improved flexibility and water resistance, Tian *et al.* [125] used waterborne polyurethane (WPU) to modify SPI. Waterborne polyurethane is an environmentally friendly substitute of polyurethane, one of the most popular groups of biomaterials applied for medical devices, with properties covering from high performance elastomer to tough thermoplastic [126]. The generated blend films exhibited fine compatibility, as a result of the strong hydrogen bonding interactions between SPI and WPU. The flexibility, water resistance, high optical transparence, and hydrophobicity of the soy protein-based films were improved significantly. Moreover, the incorporation of WPU enhanced the mechanical properties of the blend films in water, leading to the applications under wet conditions. These films also exhibited much lower cytotoxicity than WPU. The SPI/WPU blend films showed potential for application as biomaterials since they proved to be safe, biodegradable and biocompatible. In auspicious study carried out by Silva *et al.* [32], a cross-linked porous structure based on chitosan and soy protein blend were developed by means of combining a sol-gel process with the freeze-drying technique. These structures demonstrated physicochemical properties that appear to have potential to be used as a scaffold for tissue engineering of cartilage, which can be applied on cartilage implantations in some serious cases of sports injury, osteoarthritis, osteoarthritis dissecans, and infection.

Regarding the drugs delivery system in the biomedical field, Liu *et al.* [30] studied the blending process by synthesizing the soy protein/ poly (acrylic acid) forming an interpenetrating polymer network (IPN) hydrogels to investigate the possibility of this hydrogels for controlled drug release, and demonstrated that the IPN hydrogels are suitable approach for colon drug delivery. Another promising study involving the usage of soy protein-based blends was performed by Luo *et al.* [35] who prepared cellulose/ SPI sponges using a freeze-drying process. The biocompatibility and biodegradability of the blend were evaluated *in vitro* and *in vivo* by cell culture and implantation experiments. The results demonstrated that these new cellulose/SPI sponges, due to their good biocompatibility and biodegradation, can be used in biomaterials formulations such as cell and tissue scaffolds, implantation devices, and drug delivery devices.

4.3.5 Membranes

The membrane separation science has attracted much attention, since membranes have been widely applied in technologies as dialysis, ultrafiltration, fractionation of polymer mixtures, and biomedical functional membranes [34]. Certainly, the construction of membranes could be explained in the subsections already discussed in the present chapter (membranes in tissue engineering, for example) however this subject gained a special attention since represents an important topic inside the chemical technology and membranes are used in a wide range of applications.

Luo *et al.* [35] prepared a series of cellulose/SPI membranes, and observed that porous structure and the size of the pores in the surfaces increased with an increase of SPI content, and the incorporation of SPI in cellulose changed the compositions and microstructure, improving the biocompatibility of the membranes. Suggesting that biomaterials made of cellulose modified by SPI might be very promising in the biomedical fields. Santos *et al.* [122] evaluated the performance of newly developed chitosan/soy-based membranes as wound-dressing materials in a rat wound-dressing model of partial-thickness skin wounds. Chitosan/soy-based membranes promoted a faster wound repair. Besides these new membranes proved to have features regarding healing/repair stimulation, ease of handling, and final esthetic appearance, which are valuable properties for wound dressings. Silva *et al.* [124] evaluate the biocompatibility of the blend system membrane composed by chitosan and soy protein prepared by solvent casting methodology. The biological studies demonstrated a better cell adhesion of the membranes enhancing the capability of the material to support cell growth, representing an interesting result to biomedical field mainly in laboratorial membrane-based tests.

Another approach studied by Silva *et al.* [37] was investigating the cytotoxicity of soy and casein-based membranes by biochemical tests correlating with mechanical properties of these materials. The results showed that the materials demonstrated a good behavior in the presence of cells, inducing, in some cases, the cell proliferation and their surface properties seemed to be suitable for cell adhesion, which can be highly advantageous in biomedical applications, such as drug delivery and wound repair systems. Chen *et al.* [38] prepared microporous membranes by blending cellulose and SPI and obtained promising results since the blended membranes showed a certain biocompatibility, good NaOH-resistance and tensile strength in both dry and wet states which were prepared. Moreover, it was evidenced that Vero cells could grow well on the surfaces of the microporous membranes, an interesting result regarding the application in laboratorial membrane-based tests.

4.4 Applications of Soy Protein-Based Composites

In recent years, many protein-based composite materials have been studied for different applications, for example, coating, packaging, agriculture and medicine, due to the ease of processability, abundant resource, low cost, specific mechanical properties, and environmental friendliness [15–17]. The development of biodegradable composites from natural materials reduces the construction waste, the need of synthesizing petroleum-based polymers, increases energy efficiency while promoting the concept of sustainability [12, 15].

There is a large diversity of materials with varying physical and chemical properties, as the proteins, being researched for use in natural composites, including films, foams, gels, fibers, grafts, and particles [12, 16, 89, 127, 128]. Proteins display critical structural, functional and bioactive properties that have evolved in nature for millions of years [16]. It is important to note, that depending on the specific protein, there may be useful functions, such as mechanical toughness, while other critical features may be more limiting, such as higher moisture absorption, inferior fire resistance, variation in quality and price, and difficulty using established manufacturing practices when compared to synthetic composites [12, 16]. However, these characteristics can be minimized by altering the molecular structure or conformation of the protein through physical, chemical or enzymatic agents at the secondary, tertiary, and quaternary levels, such as crosslinking, blending with hydrophobic additives, acylation, enzymatic treatment, oxidation, UV irradiation, copolymerization and ultrasonic/microwave [15, 129–131].

Among a great deal of biocomposites, soy-based materials represent important source owing to their high protein content and excellent physicochemical properties which seem to offer the greatest potential for competing with synthetic polymers in terms of cost and performance [3, 12, 17]. Soy protein-based edible films, for example, have demonstrated a large potential for application due to their excellent film-forming abilities, low cost, barrier properties against oxygen, lipid and aroma permeation, homogeneous morphology, adequate tensile strength, elongation at break, higher resistance against water [14, 15, 117, 132, 133].

Ciannamea *et al.* [14] developed a biodegradable composite of soybean protein concentrate (SPC) using glycerol and dialdehyde starch as plasticizer and cross-linking agent, respectively. The soy protein-based film was used as vehicle for the delivery of *Duddingtonia flagrans* chlamydospores for the biological control of gastric nematodes in ruminals. According to Ciannamea *et al.* [14], the controllable release devices (films) used for biological control should be homogeneous, mechanically strong such it do not fracture during processing or storing but sufficiently ductile to be shaped into the desired dosage form, for example, for administration to the animal via an esophagogastric probe to the rumen where spores would be released. Their SPC films cross-linked with 5 wt% dialdehyde starch and plasticized with 30 wt% glycerol exhibited interesting physicochemical properties for application as an agent for biological control, such as homogeneous morphology, adequate tensile strength and elongation at break and swelling profile. The studies to verify the chlamydospores release *in vitro* were performed exposing the SPC films to fresh ruminal fluid and the results indicated a releasing of 4.9% for at least 24 h, suggesting that the spores release rate and duration can be better optimized by the evaluation of different SPC film formulations and process parameters, including more effective cross-linking agents and less hydrophilic plasticizers.

Lee and Min [134] developed antimicrobial films based on defatted soybean meal and the lactoperoxidase system and evaluated the application of this film as coating of ham samples. Different formulations composed by a water suspension of defatted soybean meal (10.0%) or a mixture of soybean carbohydrate, soy protein, and oil were processed with a high-shear probe mixer and then treated by one or two passes of a high-pressure homogenizer. Glycerol was evaluated as plasticizer at 10%, 20%, 30%, or 50% (w/w defatted soybean meal). Physical-mechanical properties, including color, tensile strength, percentage elongation at break, elastic modulus, and moisture-barrier properties were evaluated. Film antimicrobial activity was tested against *Salmonella typhimurium* DT 104. The results showed that the color, tensile, and moisture-barrier properties of defatted soybean meal films were affected by the concentration of glycerol plasticizer and the carbohydrate, protein, oil composition of the defatted soybean meal. However, these same parameters did not significantly affect the antimicrobial activity, which the films incorporating the lactoperoxidase system at 5% presented zones of inhibition for *S. typhimurium* DT 104 ranging from <0.5 to 1.0 mm at inoculum levels ranging from 4 log CFU/cm^2 to 2 log CFU/cm^2. The sensory properties of sliced ham, including the intensity of glossiness, odor, hardness, cohesiveness, and flavor did not significantly affect by the coating with the defatted soybean meal film, suggesting the potential for the application to commercial ham or other meat products.

Edible film can be defined as a free standing thin layer of edible material which can be applied between food components or on the surface of the food system, extending the shelf life and improve the quality of food by acting as moisture, gas, aroma, and lipid barriers. These films provide protection to a food product after the primary package is opened and help to maintain the mechanical and rheological properties, color and appearance of foods [79, 135, 136]. Edible films prepared from SPI and cysteine were tested for coating of fresh-cut "Telma" eggplants during storage under modified atmosphere packaging and some quality parameters, including color, weight loss, and firmness were studied. The cysteine content of the SPI-based coating was studied at 0.5% in a first test and at 1.0% in a second experiment. The results showed that the coated samples packed under modified atmosphere under 80 kPa O_2 and control (atmospheric conditions) resulted in the highest whiteness index values during storage, however the samples packed under a gas mixture composed by 15 kPa CO_2 + 5 kPa O_2 had the lowest whiteness index and consequently, did not improve the shelf-life of minimally processed eggplants. The storage under modified atmosphere (80 kPa O_2) and atmospheric conditions were more adequate to maintain firmness. On the other hand, the samples showed lower weight loss when stored under the two conditions of modified atmosphere compared to atmospheric conditions. For the samples packed under atmospheric conditions, the maximum commercial shelf-life was reached on day 6. Findings indicate that cysteine may be used as a natural means for improving specific properties of SPI films, extending to 9 storage days in the edible coating with 1.0% cysteine under atmospheric conditions [136].

González and Igarzabal [137] reported the preparation and characterization of biodegradable bilayer films based on isolated soy protein (SPI) and poly (lactic acid) (PLA). The effects of the incorporation of antimicrobial agents (natamycin and thymol) into SPI-PLA films on an active packaging system were examined. The results showed that the films showed interesting properties such as high transparency and strong adhesion between layers. In addition, the application of the PLA layer presented an important effect on the mechanical properties of the films, decreasing the elongation at break and increasing the tensile strength and the Young's modulus, resulting in a material less elongable and more resistant compared to those of pure SPI films. The films incorporated with an antifungal (natamycin) and an antibacterial agent (thymol) and used as active packaging showed efficient effects on growth inhibition of mold (*Aspergillus* sp.), yeast (*Saccharomyces cerevisiae*) and two strains of bacteria (*Escherichia coli* and *Staphylococcus aureus*) by *in-vitro* microbiological assays.

Soybean protein adhesives are nontoxic composites from renewable biomass that show great potential for industrial applications. It is important to note that interesting properties of protein sources, such as peptizing and protective colloid action, making them compatible with most other binders and components used in coating formulations, maximizing their use in bio-based adhesives [138, 139]. High reliability of the adhesive joint in the initial stage of drying, high adhesion strength, short setting time, pseudoplastic flow behavior and water resistance are some important properties of quality for adhesive materials [106, 138, 140]. However, the poor moisture resistance performance and low adhesion strength have been limited the use of soy protein-based adhesives in practical applications [139, 141]. Lei *et al.* [141] prepared a soy-based adhesive using defatted soy flour for plywood. Epoxy resin, melamine–formaldehyde and their mixture were used as cross-linkers to improve the water resistance of soy-based adhesive. The results indicated that the three cross-linkers presented an important role to improve the water resistance of soy-based adhesive. Soy adhesive without cross-linking showed no water resistance, reaching 100% delamination of specimens when soaked in 63 °C water for 3 h. On the other hand, the samples prepared with a mixture of epoxy resin and melamine–formaldehyde was proved to be the best cross-linker for soy-based adhesive, indicated by the absence of specimen delaminated during the soak test.

The introduction of protein-based biomaterials has been showed extreme importance in biomedical applications. Composites (scaffolds) based on soy protein modified with heat treatment and cross-linked with transglutaminase was fabricated and tested in human mesenchymal stem cells for tissue regeneration [142]. The scaffolds obtained were characterized in terms of their physico-mechanical properties and used for cultivation of human mesenchymal stem cells for evaluation of the changes in stem cell growth and morphology. The scaffold microstructure revealed rough surfaces, irregular pore shapes with sizes distributed in the range 10–125 μm, 2–5% moisture content and compressive moduli ranging between 50 Pa and 100 Pa. Enzyme treatments with transglutaminase significantly affect the moisture content and the pore sizes of the scaffolds, decreasing these values with the increasing amounts of applied enzyme. Scaffolds obtained from 3.0% SPI cross-linked with transglutaminase increased the degradation time compared to control groups (without enzyme treatment), which resulted in supported cell attachment and cell viability for up to 14 days. Scaffolds containing 5.0% SPI cross-linked with 1 U of transglutaminase were stable in media, but did not allow the cell proliferation, while 3.0% SPI scaffolds treated with the same enzyme amount promoted the cell proliferation. These results indicated that the scaffolds obtained had great potential to be used in tissue engineering applications [142].

4.5 Applications of Soy Protein-Based Nanocomposites

Nanocomposites are a class of hybrid materials divided in two parts, one with organic and another with inorganic components, which are in a nanometer scale disperse in a polymeric matrix. The interest about this subject increased due to their excellent properties when compared to the virgin and conventional polymers.

The use of natural polymers to produce nanocomposites has been increasing due to its environmental appeal to replace present petrochemical-based material. New approaches using green materials to produce bio-nanocomposites or eco-nanocomposites by incorporating organic and inorganic nanofilters, grafting polymer chains onto nanoparticle surface has been developed [6,7].

Soy protein is used in different forms such as flour, concentrate, texturized and isolate protein to reinforce some materials properties as tensile strength, elongation at break, barrier, and others mechanical performances. The possible applications of soy protein nanocomposites include adhesives, asphalts, resins, cleaning materials, paper coatings, plastics, polyesters, and textile fibers allowing their use in a large area such as packing, medical, aerospace, and automotive fields.

4.5.1 Aerospace and Automotive Structures

Initially developed for aerospace structures, fiber-reinforced "green" nanocomposites are now found in applications from automotive parts to high-performance sporting goods where their properties per unit mass are critical for their performance. Both the Boeing 787 "Dreamliner" and Airbus A380 used a significant amount of these nanocomposites in their structure to reduce their mass and increase the fuel efficiency and payload [20].

The use of nanocomposites in automotive body panels is an interesting perspective of saving cost and decrease the environmental impact when compared with the use of steel and aluminum [143].

To produce automotive panels just moderate properties are required, high strength and stiffness are not necessary, thus allowing the use of plant-based cellulosic fibers to build these nanocomposites [144, 145].

In the study of Huang *et al.* [146], self-assembled rigid supramolecular nanoplatelets from pluronic polymers with many lengths of polyethylene oxide and β-cyclodextrin were used to reinforce soy protein-based plastics. The β-ciclodextrin powder was mixed with the pluronic polymer to produce a mixing solution in which a molar ratio between both was controlled. The authors observed that only a small amount of nanoplateles was able to disperse uniformly into SPI matrix and with the increasing of the nanoplatelets amount lead to a decreasing of mechanical performance due to a formation of rectangle objects into the matrix.

Nakamura *et al.* [20] developed a nanocomposite based on soy protein concentrate modified with glycerol and/or halloysite nanotubes. These authors characterized extensively the nanocomposites properties and mechanical performances to evaluate their use in vehicle and infrastructure applications. Promising results were reported; soy protein concentrate has low flammability when compared with petrochemical resins, such as epoxies and vinyl esters. Another interesting aspect is that the incorporation of 5% mass fraction of halloysites nanotubes (HNT) decrease the flammability without affects the mechanical properties.

4.5.2 Packing

Packing films and coatings used on pharmaceuticals, confections, fruits, vegetables, and some meat products, are usually from lipids, proteins, carbohydrates, or composites of the three. They provide extra protection against desiccation, oxidation, or other modes of deterioration. Interest in protein-based films has been centered on casein, collagen, corn zein, gelatin, soy protein, and wheat gluten [19].

The properties of forming brilliant film made of soy protein a potential biopolymer source to use as packing material, however, mechanical and barrier properties should be improved as long as soy-based films are brittle being necessary add plasticizers to obtain flexibility [147]. To solve this problem one possible technique is reinforce the soy protein polymer matrix with montmorillonite (MMT).

Kumar *et al.* [148] prepared films based on SPI and with MMT using melt extrusion. The authors achieved important improvement in mechanical (tensile strength and percent elongation at break) and dynamic mechanical properties (glass transition temperature and storage modulus), thermal stability and water vapor permeability when reinforced these material. Among the variables investigated, pH of films forming solutions, MMT content and extrusion conditions were the variables that affected more the films properties. The resultant film made of soy/MMT is recommended to be used for packing food with high moisture content as fresh fruit and vegetable in order to replace low density polyethylene (LDPE) and polyvinylidene chloride (PVDC).

4.5.3 Medical

Nanocomposite materials involving electrospun fibers are of particular interest, and probably one of the most promising application areas for modern electrospinning is biomedical materials. Auspicious applications of electrospun fibers in biomedical engineering include tissue-engineering scaffolds, controlled drug release depots, wound dressings, coatings for medical-implant devices, nanocomposite materials for dental restoration, membranes for molecular separations, and biosensor coatings [149].

Regarding the usage of soy protein, Vega-Lugo and Lim [31] studied electrospun fibers from SPI/poly (ethylene oxide) (PEO) blend and poly (lactic acid) for the encapsulation of allylisothiocyanate (AITC), a natural occurring antimicrobial compound, and among other results observed that the release of AITC increased proportionally as relative humidity increased, revealing a antimicrobial nanofiber which may be useful in active packing applications.

A promising work was carried out by Reddy and Yang [150] which produced soy protein fibers using the wet spinning method and evidenced that those fibers have wet strength of 19 MPa compared with 1.2–2.4 MPa for noncross-linked collagen fibers. The authors also reported that fibroblasts grown on soy protein fibers showed good attachment, growth, and metabolic activities indicating that those fibers are suitable as tissue engineering substrates.

4.6 Conclusion

It has been demonstrated that soy protein, an abundant and cheap raw material, showed good potential for development of biomaterials with several applications. New techniques, including physical, chemical and enzymatic methods, have been developed to build soy protein-based blends, composites and nanocomposites and their interactions with different components to enhance specific mechanical properties. Some applications include the use of these biomaterials as components of adhesives, films, resins, paper coatings and plastics in particular areas such as biomedical, optoelectronic, optical coating and packaging. It is important to note that most of these studies were carried out on a laboratory scale, so there is a lack of data about industrial application in large scale. The use of soy protein in large scale would be very important for the environment issue as long as is a biodegradable material. It is essential that future research in the field of biomaterials using soy proteinis focused to improve important characteristics for application, such as mechanical and thermal properties, durability and moisture resistance to make them competitive with the synthetic materials.

References

1. USDA, United States Department of Agriculture. http://soystats.com/international-world-soybean-production/
2. Egbert, W.R., *Isolated soy protein: technology, properties, and applications.* AOCS Press, 134–162, 2004.
3. Mohanty, A.K., Liu, W., Tummala, P., Drzal, L.T., Misra, M., Narayan, R., Soy protein-based plastics, blends, and composites. In: *Natural fibers, biopolymers, and biocomposites*, A.K. Mohanty, M. Misra, L.T. Drzal, Eds., CRC Press, 2005.
4. Yang, S., Leong, K.F., Du, Z., Chua, C.K., The design of scaffolds for use in tissue engineering. Part I. Traditional factors. *Tissue Eng.*, 7, 679–689, 2001.
5. Vaz, C.M., Van Doeveren, P.F., Reis, R.L., Cunha, A.M., Soy matrix drug delivery systems obtained by melt-processing techniques. *Biomacromolecules*, 4, 1520–1529, 2003a.
6. Angellier, H., Molina-Boisseau, S., Dole, P., Dufresne, A., Thermoplastic starch-waxy maize starch nanocrystals nanocomposites. *Biomacromolecules*, 7, 531–539, 2006.
7. *Chen, P., Zhang,* L., Interaction and properties of highly exfoliated soy protein/montmorillonite nanocomposites. *Biomacromolecules*, 7, 1700–1706, 2006.
8. Regalado, C., Perez-Perez, C., Lara-Cortes, E., Garcia-Almendarez, B., Guevara-González, R.G., Torres-Pacheco, I., Whey protein based edible food packaging films and coatings. *Adv. Agric. Food Biotechnol.*,237–261, 2006.
9. Senyurt, A.F., Warren, G., Whitehead, J.B. Jr, Hoyle, C.E., Matrix physical structure effect on the electro-optic characteristics of thiol-ene based H-PDLC films. *Polymer*, 47,2741–2749, 2006.
10. Li, Y.D., Zeng, J.B., Wang, X.L., Yang, K.K., Wang, Y.Z., Structure and properties of soy protein/poly (butylene succinate) blends with improved compatibility. *Biomacromolecules*, 9, 3157–3164, 2008.
11. Chen, Z., Chisholm, B.J., Patani, R., Wu, J.F., Fernando, S., Jogodzinski, K., Webster, D.C., Soy-based UV-curable thiol-ene coatings. *J. Coat. Technol. Res.*, 7, 603–613, 2010.
12. Dittenber, D.B., Ganga Rao, H.V.S., Critical review of recent publications on use of natural composites in infrastructure. *Composites: Part A*, 43, 1419–1429, 2012.
13. Zhu, R., Liu, H., Zhang, J., Compatibilizing effects of maleated poly (lactic acid) (PLA) on properties of PLA/Soy protein composites. *Ind. Eng. Chem. Res.*, 51,7786–7792, 2012.

14. Ciannamea, E.M., Sagüés, M.F., Saumell, C., Stefani, P.M., Ruseckaite, R.A., Soybean protein films. Characterization and potential as novel delivery devices of *Duddingtoniaflagrans* chlamydospores. *Biolog. Contr.*, 66, 92–101, 2013.
15. González, A., Strumia, M.C., Igarzabal, C.I.A., Cross-linked soy protein as material for biodegradable films: synthesis, characterization and biodegradation. *J. Food Eng.*, 106, 331–338, 2011.
16. Hu, X., Cebe, P., Weiss, A.S., Omenetto, F., Kaplan, D.L., Protein-based composite materials. *Materials Today*, 15,208–215, 2012.
17. Xie, D.-Y., Son, F., Zhang, M., Wang, X.-L., Wang, Y.-Z., Soy protein isolate films with improved property via a facile surface coating. *Ind. Crops Prod.*, 54, 102–108, 2014.
18. Ji, J., Li, B., Zhong, W.H., An ultraelastic poly (ethylene oxide)/soy protein film with fully amorphous structure. *Macromolecules*, 45, 602–606, 2011.
19. Brandenburg, A.H., Weller, C.L., Testin, R.F., Edible films and coatings from soy protein. *J. Food Sci.*, 58, 1086–1089, 1993.
20. Nakamura, R., Netravali, A.N., Morgan, A.B., Nyden, M.R., Gilman, J.W., Effect of halloysite nanotubes on mechanical properties and flammability of soy protein based green composites. *Fire Mater.*, 37, 75–90, 2013.
21. Peles, Z., Zilberman, M., Novel soy protein wound dressings with controlled antibiotic release: mechanical and physical properties. *Acta Biomaterialia*, 8, 209–217, 2012.
22. Tian, H., Wang, Y., Zhang, L., Quan, C., Zhang, X., Improved flexibility and water resistance of soy protein thermoplastics containing waterborne polyurethane. *Ind. Crops Prod.*, 32, 13–20, 2010a.
23. Qiu, Y., Park, K., Environment-sensitive hydrogels for drug delivery. *Adv. Drug Deliv. Rev.*, 53, 321–339, 2001.
24. Chen, L., Remondetto, G., Rouabhia, M., Subirade, M., Kinetics of the breakdown of cross-linked soy protein films for drug delivery. *Biomaterials*, 29, 3750–3756, 2008.
25. Vaz, C.M., Van Doeveren, P.F.N.M., Reis, R.L., Cunha, A.M., Development and design of double-layer co-injection moulded soy protein based drug delivery devices. *Polymer*, 44, 5983–5992, 2003b.
26. Maltais, A., Remondetto, G.E., Subirade, M., Soy protein cold-set hydrogels as controlled delivery devices for nutraceutical compounds. *Food Hydrocolloid.*, 23, 1647–1653, 2009.
27. Meikle, S.T., Standen, G., Salvage, J., De Santis, R., Nicolais, L., Ambrosio, L., Santin, M., Synthesis and characterization of soybean-based hydrogels with an intrinsic activity on cell differentiation. *Tissue Eng. Part A*, 18, 1932–1939, 2012.
28. Vaz, C.M., Fossen, M., Van Tuil, R.F., De Graaf, L.A., Reis, R.L., Cunha, A.M., Casein and soybean protein-based thermoplastics and composites as alternative biodegradable polymers for biomedical applications. *J. Biomed. Mater. Res. Part A*, 65, 60–70, 2003c.

29. Swetha, M., Sahithi, K., Moorthi, A., Srinivasan, N., Ramasamy, K., Selvamurugan, N., Biocomposites containing natural polymers and hydroxyapatite for bone tissue engineering. *Int. J. Biol. Macromol.* 47, 1–4, 2010.
30. Liu, Y., Cui, Y., Yin, G., Ma, H., Synthesis, characterization, and drug release behaviour of novel soy protein/poly (acrylic acid) IPN hydrogels. *Iranian Polym. J.*, 18,339–348, 2009.
31. Vega-Lugo, A.C., Lim, L.T., Controlled release of allylisothiocyanate using soy protein and poly (lactic acid) electrospun fibers. *Food Res. Int.*, 42, 933–940, 2009.
32. Silva, S.S., Oliveira, J.M., Mano, J.F., Reis, R.L., Physicochemical characterization of novel chitosan-soy protein/TEOS porous hybrids for tissue engineering applications. *Mater. Sci. Forum,* 514, 1000–1004, 2006.
33. Luo, L.H., Zhang, Y.F., Wang, X.M., Wan, Y.Q., Chang, P.C., Anderson, D.R., Chen, Y., Preparation, characterization, and *in vitro* and *in vivo* evaluation of cellulose/soy protein isolate composite sponges. *J. Biomater. Appl.*, 24, 503–526, 2010.
34. Snyders, R., Shingel, K.I., Zabeida, O., Roberge, C., Faure, M.P., Martinu, L., Klemberg-Sapieha, J.E., Mechanical and microstructural properties of hybrid poly (ethylene glycol)–soy protein hydrogels for wound dressing applications. *J. Biomed. Mater. Res. Part A*, 83, 88–97, 2007.
35. Luo, L.H., Wang, X.M., Zhang, Y.F., Liu, Y.M., Chang, P.R., Wang, Y., Chen, Y., Physical properties and biocompatibility of cellulose/soy protein isolate membranes coagulated from acetic aqueous solution. *J. Biomater. Sci., Polym. Ed.*, 19, 479–496, 2008.
36. Silva, R.M., Elvira, C., Mano, J.F., San Roman, J., Reis, R.L., Influence of β-radiation sterilization in properties of new chitosan/soybean protein isolate membranes for guided bone regeneration. *J. Mater. Sci.: Mater. Med.* 15, 523–528, 2004.
37. Silva, G.A., Vaz, C.M., Coutinho, O.P., Cunha, A.M., Reis, R.L., *In vitro* degradation and cytocompatibility evaluation of novel soy and sodium caseinate-based membrane biomaterials. *J. Mater. Sci.: Mater. Med.*, 14, 1055–1066, 2003.
38. Chen, Y., Zhang, L., Gu, J., Liu, J., Physical properties of microporous membranes prepared by hydrolyzing cellulose/soy protein blends. *J. Membr. Sci.*, 241, 393–402, 2004.
39. Wang, Y., Cao, X., Zhang, L., Effects of cellulose whiskers on properties of soy protein thermoplastics. *Macromol. Biosci.*, 6, 524–531, 2006.
40. Samir, M.A.S.A., Alloin, F., Dufresne, A., Review of recent research into cellulosic whiskers, their properties and their application in nanocomposite field. *Biomacromolecules*, 6, 612–626, 2005.
41. Lu, Y., Weng, L., Zhang, L., Morphology and properties of soy protein isolate thermoplastics reinforced with chitin whiskers. *Biomacromolecules*, 5, 1046–1051, 2004.

42. Demus, D., Goodby, J., Gray, G.W., Spiess, H.W., Vill, V., *Handbook of liquid crystals*, Vol. 1: *Fundamentals*. Wiley VCH, 2008.
43. Lagerwall, J.P., Schütz, C., Salajkova, M., Noh, J., Park, J.H., Scalia, G., Bergström, L., Cellulose nanocrystal-based materials: from liquid crystal self-assembly and glass formation to multifunctional thin films. *NPG Asia Mater.*, 6, 1–12, 2014.
44. Godbout, D.L., Gray, D.G., Revol, J.F.,U.S. Patent No. 5,629,055. Washington, DC: U.S. Patent and Trademark Office, 1997.
45. Werbowyj, R.S., Gray, D.G., Liquid crystalline structure in aqueous hydroxypropyl cellulose solutions. *Mol. Cryst. Liq. Cryst.*, 34, 97–103, 1976.
46. Werbowyj, R.S., Gray, D.G., Optical properties of hydroxypropyl cellulose liquid crystals. I. Cholesteric pitch and polymer concentration. *Macromolecules*, 17, 1512–1520, 1984.
47. Gray, D.G., Chemical characteristics of cellulosic liquid crystals. *Faraday Discuss. Chem. Soc.*, 79, 257–264, 1985.
48. Gilbert, R.D., Patton, P.A., Liquid crystal formation in cellulose and cellulose derivatives. *Prog. Polym. Sci.*, 9, 115–131, 1983.
49. Bhadani, S.N., Gray, D.G., Cellulose-based liquid crystalline polymers; esters of (hydroxypropyl) cellulose. *Mol. Cryst. Liq. Cryst.*, 99, 29–38, 1983.
50. Angles, M.N., Dufresne, A., Plasticized starch/tunicin whiskers nanocomposites. 1. Structural analysis. *Macromolecules*, 33, 8344–8353, 2000.
51. Scrosati, B., Garche, J., Lithium batteries: Status, prospects and future. *J. Power Sources*, 195, 2419–2430, 2010.
52. Samir, M.A.S.A., Alloin, F., Sanchez, J.Y., El Kissi, N., Dufresne, A., Preparation of cellulose whiskers reinforced nanocomposites from an organic medium suspension. *Macromolecules*, 37, 1386–1393, 2004.
53. Lalia, B.S., Samad, Y.A., Hashaikeh, R., Nanocrystalline cellulose-reinforced composite mats for lithium-ion batteries: electrochemical and thermomechanical performance. *J. Solid State Electrochem.*, 17, 575–581, 2013.
54. Chiappone, A., Nair, J.R., Gerbaldi, C., Jabbour, L., Bongiovanni, R., Zeno, E., Beneventi, D., Penazzi, N., Microfibrillated cellulose as reinforcement for Li-ion battery polymer electrolytes with excellent mechanical stability. *J. Power Sources*, 196, 10280–10288, 2011.
55. Jabbour, L., Bongiovanni, R., Chaussy, D., Gerbaldi, C., Beneventi, D., Cellulose-based Li-ion batteries: a review. *Cellulose*, 20, 1523–1545, 2013.
56. Chun, S.J., Choi, E.S., Lee, E.H., Kim, J.H., Lee, S.Y., Lee, S.Y., Eco-friendly cellulose nanofiber paper-derived separator membranes featuring tunable nanoporous network channels for lithium-ion batteries. *J. Mater. Chem.*, 22, 16618–16626, 2012.
57. Baumeister, P., *Optical coating technology*. Bellingham: SPIE Press, 137, 2004.
58. Lee, C.C., Thin film optical coatings. In: *The current trends of optics and photonics*. Springer Netherlands, 129, 3–33, 2015.
59. Teng, G., Soucek, M.D., Epoxidized soybean oil-based ceramer coatings. *J. Am. Oil Chem. Soc.*, 77, 381–387, 2000.

60. Bajpai, M., Shukla, V., Singh, D.K., Singh, M., Shukla, R., A study of the film properties of pigmented UV-curable epoxidised soybean oil. *Pigm. Resin Technol.*, 33, 160–164, 2004.
61. Cai, J., Shu, W., Yun, L., Jiu, L., Influence of reactive diluents on UV curing acrylic epoxidized soybean oil system. *Tuliao Gongye*, 36, 12–15, 2006.
62. Lee, S.H., Lee, S.Y., Lim, H., Nam, J., Kye, H., Lee, Y., Preparation and properties of plasticized cellulose diacetate using triacetine/epoxidized soybean oil. *Polymer (Korea)*, 30, 202–206, 2006.
63. Ji, J.Y., Lively, B., Zhong, W.H., Soy protein-assisted dispersion of carbon nanotubes in a polymer matrix. *Mater. Exp.*, 2, 76–82, 2012.
64. Van de Velde, K., Kiekens, P., Biopolymers: overview of several properties and consequences on their applications. *Polym. Test.*, 2, 433–442, 2002.
65. Abugoch, L.E., Tapia, C., Villamán, M.C., Yazdani-Pedram, M., Díaz-Dosque, M., Characterization of quinoa protein-chitosan blend edible films. *Food Hydrocolloid.*, 25, 879–886, 2011.
66. Bigi, A., Cojazzi, G., Panzavolta, S., Rubini, K., Mechanical and thermal properties of gelatin films at different degrees of glutaraldehyde cross-linking. *Biomaterials*, 22, 763–768, 2001.
67. González, A., Igarzabal, C.I.A., Nanocrystal-reinforced soy protein films and their application as active packaging. *Food Hydrocolloid.*, 43, 777–784, 2015.
68. Guerrero, P., de la Caba, K., Thermal and mechanical properties of soy protein films processed at different pH by compression. *J. Food Eng.*, 100, 261–269, 2010.
69. Ciannamea, E.M., Stefani, P.M., Ruseckaite, R.A., Physical and mechanical properties of compression molded and solution casting soybean protein concentrate based films. *Food Hydrocolloid.*, 38, 193–204, 2014.
70. Park, S.K., Hettiarachchy, N.S., Ju, Z.Y., Gennadios, A., Formation and properties of soy protein films and coatings In: *Protein-based films and coatings*, A. Gennadios, Ed., pp. 123–137. Boca Raton, FL: CRC Press, 2002.
71. Gennadios, A., Brandenburg, A.H., Weller, C.L., Testin, R.F., Effect of pH on properties of wheat gluten and soy protein isolate films. *J. Agric. Food Chem.*, 41, 1835–1839, 1993.
72. Gennadios, A., Weller, C.L., Edible films and coatings from soymilk and soy protein. *Cereal Foods World*, 36, 1004–1009, 1991.
73. Cho, S.Y., Rhee, C., Mechanical properties and water vapor permeability of edible films made from fractionated soy proteins with ultrafiltration. *LWT - Food Sci. Technol.*, 37, 833–839, 2004.
74. Kokoszka, S., Debeaufort, F., Hambleton, A., Lenart, A., Voilley, A., Protein and glycerol contents affect physico-chemical properties of soy protein isolate-based edible films. *Innov. Food Sci. Emerg. Technol.*, 11, 503–510, 2010.
75. Guilbert, S., Gontard, N., Cuq, B., Technology and applications of edible protective films. *Packag. Technol. Sci.*, 8, 339–346, 1995.
76. Arancibia, M.Y., López-Caballero, M.E., Gómez-Guillén, M.C., Montero, P., Release of volatile compounds and biodegradability of active soy protein

lignin blend films with added citronella essential oil. *Food Contr.*, 44, 7–15, 2014.
77. Piazza, L., Durr-Auster, N., Gigli, J., Windhab, E.J., Fischer, P., Interfacial rheology of soy proteins – High methoxyl pectin films. *Food Hydrocolloid.*, 23, 2125–2131, 2009.
78. Zhang, J., Mungara, P., Jan, J., Mechanical and thermal properties of extruded soy protein sheets. *Polymer*, 42, 2569–2578, 2001.
79. Cao, N., Fu, Y., He, J., Preparation and physical properties of soy protein isolate and gelatin composite films. *Food Hydrocolloid.*, 21, 1153–1162, 2007.
80. Entwistle, C.A., Rowe, R.C., Plasticization of cellulose ethers used in the film coating of tablets. *J. Pharm. Pharmacol.*, 31, 269–272, 1979.
81. Banker, G.S., Film coating theory and practice. *J. Pharm. Sci.*, 55, 81–89, 1966.
82. Guilbert, S., *Technology and application of edible protective films in food packaging and preservation: theory and practice. Elsevier*,71–394, 1986.
83. Hernandez-Izquierdo, V.M., Krochta, J.M., Thermoplastic processing of proteins for film formation- a review. *J. Food Sci.*, 73, 30–39, 2008.
84. Hettiarachchy, N.S., Eswaranandam, S., Edible films and coatings from soybean and other protein sources. In: *Bailey's industrial oil and fat products.* John Wiley and Sons, Inc, 2005.
85. McHugh, T.H., Krochta, J.M., Sorbitol- vs glycerol-plasticized whey protein edible films: integrated oxygen permeability and tensile property evaluation. *J. Agric. Food Chem.*, 42, 841–845, 1994.
86. Rhim, J.W., Gennadios, A., Handa, A., Weller, C.L., Hanna, M.A., Solubility, tensile, and color properties of modified soy protein isolate films. *J. Agric. Food Chem.*, 48, 4937–4941, 2000.
87. Singh, P., Kumar, R., Sabapathy, S.N., Bawa, A.S., Functional and edible uses of soy protein products. *Compr. Rev. Food Sci. Food Saf.*, 7, 14–28, 2008.
88. Sothornvit, R., Krochta, J.M., Plasticizer effect on mechanical properties of [beta]-lactoglobulin films. *J. Food Eng.*, 50, 149–155, 2001.
89. Guerrero, P., Stefani, P.M., Ruseckaite, R.A., de la Caba, K., Functional properties of films based on soy protein isolate and gelatin processed by compression. *J. Food Eng.*, 105, 65–72, 2011a.
90. Mo, X., Sun, X., Plasticization of soy protein polymer by polyol-based plasticizers. *J. Am. Oil Chem. Soc.*, 79, 197–202, 2002.
91. Ustunol, Z., Mert, B., Water solubility, mechanical, barrier, and thermal properties of cross-linked whey protein isolate-based films. *J. Food Sci.*, 69, 129–133, 2004.
92. Guerrero, P., Nur Hanani, Z.A., Kerry, J.P., de la Caba, K., Characterization of soy protein-based films prepared with acids and oils by compression. *J. Food Eng.*, 107, 41–49, 2011b.
93. Choi, W.Y., Lee, C.M., Park, H.J., Development of biodegradable hot-melt adhesive based on poly-e-caprolactone and soy protein isolate for food packaging system. *LWT - Food Sci. Technol.*, 39, 591–597, 2006.

94. Wang, C., Carriere, C.J., Willett, J.L., Processing, mechanical properties, and fracture behavior of cereal protein/poly (hydroxyl ester ether) blends. *J. Polym. Sci. Part B: Polym. Phys.*, 40, 2324–2332, 2002.
95. Mo, X., Hu, J., Sun, X., Ratto, J.A., Compression and tensile strength of low-density straw-protein particle board. *Ind. Crops Prod.*, 14, 1–9, 2001.
96. Su, J.F., Yuan, X.Y., Huang, Z., Xia, W.L., Properties stability and biodegradation behaviors of soy protein isolate/poly (vinyl alcohol) blend films. *Polym. Degrad. Stab.*, 95, 1226–1237, 2010.
97. Song, F., Tang, D.L., Wang, X.L., Wang, Y.Z., Biodegradable soy protein isolate-based materials: a review. *Biomacromolecules*, 12,3369–3380, 2011.
98. Park, S.K., Hettiarachchy, N.S., Were, L., Degradation behavior of soy protein-wheat gluten films in simulated soil conditions. *J. Agric. Food Chem.*, 48, 3027–3031, 2000.
99. Kakizawa, Y., U.S. Patent No. 5,844,066. Washington, DC: U.S. Patent and Trademark Office, 1998.
100. Lipinsky, E.S., Sinclair, R.G., Is lactic acid a commodity chemical. *Chem. Eng. Prog.*, 82, 26–32, 1986.
101. Kim, J.T., Netravali, A.N., Physical properties of biodegradable films of soy protein concentrate/gelling agent blends. *Macromol. Mater. Eng.*, 297, 176–183, 2012.
102. Liu, B., Jiang, L., Zhang, J., Study of effects of processing aids on properties of poly (lactic acid)/soy protein blends. *J. Polym. Environ.*, 19, 239–247, 2011.
103. Li, K., Peshkova, S., Geng, X., Investigation of soy protein-Kymene® adhesive systems for wood composites. *J. Am. Oil Chem. Soc.*, 81, 487–491, 2004.
104. Sun, X., Bian, K., Shear strength and water resistance of modified soy protein adhesives. *J. Am. Oil Chem. Soc.*, 76, 977–980, 1999.
105. Xiao, Z., Li, Y., Wu, X., Qi, G., Li, N., Zhang, K., Wang, D., Sun, X.S., Utilization of sorghum lignin to improve adhesion strength of soy protein adhesives on wood veneer. *Ind. Crops Prod.*, 50, 501–509, 2013.
106. Kalapathy, U., Hettiarachchy, N.S., Myers, D., Hanna, M.A.,Modification of soy proteins and their adhesive properties on woods. *Journal of the American Oil Chemists' Society*, 72, 507–510, 1995.
107. Amaral-Labat, G.A., Pizzi, A., Goncalves, A.R., Celzard, A., Rigolet, S., Rocha, G.J.M., Environment-friendly soy flour-based resins without formaldehyde. *J. Appl. Polym. Sci.*, 108, 624–632, 2008.
108. Steele, P.H., Kreibich, R.E., Steynberg, P.J., Hemingway, R.W., Finger jointing green southern yellow pine with a soy-based adhesives. *Adhes*, 41, 49–56, 1998.
109. Zhong, Z., Sun, X.S., Wang, D., Isoelectric pH of polyamide–epichlorohydrin modified soy protein improved water resistance and adhesion properties. *J. Appl. Polym. Sci.*, 103, 2261–2270, 2007.
110. Thames, S.F., Cook, R.C., Mendon, S.K., U.S. Patent No. 6,790,271. Washington, DC: U.S. Patent and Trademark Office, 2004.

111. Toledano, A., Serrano, L., Garcia, A., Mondragon, I., Labidi, J., Comparative study of lignin fractionation by ultrafiltration and selective precipitation. *Chem. Eng. J.*, 157, 93–99.
112. Campos, C.A., Gerschenson, L.N., Flores, S.K., Development of edible films and coatings with antimicrobial activity. *Food and Bioprocess Technology*, 4, 849–875, 2011.
113. Chinma, C.E., Ariahu, C.C., Abu, J.O., Development and characterization of cassava starch and soy protein concentrate based edible films. *Int. J. Food Sci. Technol.*, 47, 383–389, 2012.
114. Wang, L., Liu, L., Holmes, J., Huang, J., Kerry, J.F., Kerry, J.P., Effect of pH and addition of corn oil on the properties of whey protein isolate-based films using response surface methodology. *Int. J. Food Sci. Technol.*, 43, 787–796, 2008.
115. John, J., Bhattacharya, M., Properties of reactively blended soy protein and modified polyesters. *Polym. Int.*, 48, 1165–1172, 1999.
116. Monedero, F.M., Fabra, M.J., Talens, P., Chiralt, A., Effect of calcium and sodium caseinates on physical characteristics of soy protein isolate–lipid films. *J. Food Eng.*, 97, 228–234, 2010.
117. Su, J.F., Huang, Z., Liu, K., Fu, L.L., Liu, H.R., Mechanical properties, biodegradation and water vapor permeability of blend films of soy protein isolate and poly (vinyl alcohol) compatibilized by glycerol. *Polym. Bull.*, 58, 913–921, 2007.
118. Rhim, J.W., Wu, Y., Weller, C.L., Schnepf, M., Physical characteristics of emulsified soy protein-fatty acid composite films. *Sciences des Aliments*, 19, 57–71, 1999.
119. Togrul, H., Arslan, N., Carboxymethyl cellulose from sugar beet pulp cellulose as a hydrophilic polymer in coating of mandarin. *J. Food Eng.*, 62, 271–279, 2004.
120. Fabra, M.J., Talens, P., Chiralt, A., Tensile properties and water vapor permeability of sodium caseinate films containing oleic acid–beeswax mixtures. *J. Food Eng.*, 85, 393–400, 2008.
121. Santos, T.C., Höring, B., Reise, K., Marques, A.P., Silva, S.S., Oliveira, J.M., Mano, J.F., Gastro, A.G., Reis, R.L., van Griensven, M., *In vivo* performance of chitosan/soy-based membranes as wound-dressing devices for acute skin wounds. *Tissue Eng. Part A*, 19, 860–869, 2013.
122. Zhou, Z., Zheng, H., Wei, M., Huang, J., Chen, Y., Structure and mechanical properties of cellulose derivatives/soy protein isolate blends. *J. Appl. Polym. Sci.*, 107, 3267–3274, 2008.
123. Silva, S.S., Santos, M.I., Coutinho, O.P., Mano, J.F., Reis, R.L.,Physical properties and biocompatibility of chitosan/soy blended membranes. *J. Mater. Sci.: Mater. Med.*, 16, 575–579, 2005.
124. Tian, K., Shao, Z., Chen, X., Natural electroactive hydrogel from soy protein isolation. *Biomacromolecules*, 11, 3638–3643, 2010b.

125. Yoon, P.J., Han, C.D., Effect of thermal history on the rheological behavior of thermoplastic polyurethanes. *Macromolecules*, 33, 2171–2183, 2000.
126. Song, X., Zhou, C., Fu, F., Chen, Z., Wu, Q., Effect of high-pressure homogenization on particle size and film properties of soy protein isolate. *Ind. Crops Prod.*, 43, 538–544, 2013.
127. Qi, G., Li, N., Wang, D., Sun, X.S., Physicochemical properties of soy protein adhesives modified by 2-octen-1-ylsuccinic anhydride. *Ind. Crops Prod.*, 46, 165–172, 2013.
128. Kumar, R., Choudhary, V., Mishra, S., Varma, I.K., Mattiason, B., Adhesives and plastics based on soy protein products. *Ind. Crops Prod.*, 16, 155–172, 2002.
129. Wihodo, M., Moraru, C.I., Physical and chemical methods used to enhance the structure and mechanical properties of protein films: a review. *J. Food Eng.*, 114, 292–302, 2013.
130. Wang, Z., Zhang, N., Wang, H., Sui, S., Sun, X., Ma, Z., The effects of ultrasonic/microwave assisted treatment on the properties of soy protein isolate/titanium dioxide films. *LWT - Food Sci. Technol.*, 57, 548–555, 2014.
131. Kim, K.M., Weller, C.L., Hanna, M.A., Gennadios, A., Heat curing of soy protein films at selected temperatures and pressures. *LWT - Food Sci. Technol.*, 35, 140–145, 2002.
132. Garrido, T., Etxabide, A., Peñalba, M., de la Caba, K., Guerrero, P., Preparation and characterization of soy protein thin films: processing–properties correlation. *Mater. Lett.*, 105, 110–112, 2013.
133. Lee, H., Min, S.C., Antimicrobial edible defatted soybean meal-based films incorporating the lactoperoxidase system. *LWT - Food Sci. Technol.*, 54, 42–50, 2013.
134. Zhang, L., Zeng, M., *Proteins as sources of materials in monomers, polymers and composites from renewable resource*, M.N. Belgacem, A. Gandini, Eds., Elsevier, pp. 472–499, 2008.
135. Ghidelli, C., Mateos, M., Rojas-Argudo, C., Pérez-Gago, M.B., Extending the shelf life of fresh-cut eggplant with a soy protein–cysteine based edible coating and modified atmosphere packaging. *Postharv. Biol. Technol.*, 95, 81–87, 2014.
136. Lacroix, M., Vu, K.D., Edible coating and film materials: proteins. In: *Innovations in food packaging*, J.H. Han, Ed., Academic Press, San Diego, CA, USA, pp. 277–294, 2014.
137. González, A., Igarzabal, C.I.A., Soy protein e poly (lactic acid) bilayer films as biodegradable material for active food packaging. *Food Hydrocolloid.*, 33, 289–296, 2013.
138. Qi, G., Sun, X.S., Peel adhesion properties of modified soy protein adhesive on a glass panel. *Ind. Crops Prod.*, 32, 1659–1688, 2010.
139. Lin, Q., Chen, N., Bian, L., Fan, M., Development and mechanism characterization of high performance soy-based bio-adhesives. *Int. J. Adhes. Adhes.*, 34, 11–16, 2012.

140. Klimets, T.G., Grinyuk, E.V., Nareiko, E.I., Skakovskii, E.D., Tychinskaya, L.Y., Krul, L.P., Aqueous adhesives based dextrin-containing polyacrylate composites. *Russian J. Appl. Chem.*, 77, 822–824, 2004.
141. Lei, H., Du, G., Wu, Z., Xi, X., Dong, Z., Cross-linked soy-based wood adhesives for plywood. *Int. J. Adhes. Adhes.*, 50, 199–203, 2014.
142. Chien, K.B., Shah, R.N., Novel soy protein scaffolds for tissue regeneration: Material characterization and interaction with human mesenchymal stem cells. *Acta Biomaterialia*, 8, 694–703, 2012.
143. Lloyd, S., Lave, L., Life cycle economic and environmental implications of using nanocomposites in automobiles. *Environ. Sci. Technol.*, 37, 3458–3466, 2003.
144. Peijs, T., Composites turn green!. *E-Polymers*, 2, 1–13, 2002.
145. Netravali, A.N., Chabba, S., Composites get greener. *Mater. Today*, 6, 22–29, 2003.
146. Huang, J., Zhou, Z., Wei, M., Chen, Y., Chang, P.R., Soy protein-based nanocomposites reinforced by supramolecular nanoplatelets assembled from pluronic polymers/β-cyclodextrin pseudo polyrotaxanes. *J. Appl. Polym. Sci.*, 107, 409–417, 2008.
147. Wang, S., Sue, H.J., Jane, J., Effects of polyhydric alcohols on the mechanical properties of soy protein plastics. *J. Macromol. Sci.*, Part A, 33, 557–569, 1996.
148. Kumar, P., Sandeep, K.P., Alavi, S., Truong, V.D., Gorga, R.E., Preparation and characterization of bio-nanocomposite films based on soy protein isolate and montmorillonite using melt extrusion. *J. Food Eng.*, 100, 480–489, 2010.
149. Khadka, D.B., Haynie, D.T., Protein-and peptide-based electrospun nanofibers in medical biomaterials. *Nanomed.: Nanotechnol. Biol. Med.*, 8, 1242–1262, 2012.
150. Reddy, N., Yang, Y., Soyprotein fibers with high strength and water stability for potential medical applications. *Biotechnol. Prog.*, 25, 1796–1802, 2009.

5

Biomedical Applications of Soy Protein

Blessing A. Aderibigbe

Department of Chemistry, University of Fort Hare, Eastern Cape, South Africa

Abstract

Soy protein is a readily available biopolymer that has been reported to be useful in wound dressing, tissue engineering, and drug delivery systems. Its biomedical applications are attributed to its unique characteristics such as biodegradability, biocompatibility, affordability, and low immunogenicity. It is limited by poor mechanical properties and is combined with synthetic polymers and other materials resulting in improved mechanical properties. This chapter will be focused on the biomedical applications of soy protein in drug delivery systems, wound dressing, and tissue regenerative medicine.

***Keywords*:** Soy protein, wound dressing, drug delivery systems, tissue regenerative medicine

5.1 Introduction

Soy proteins (SPs) are natural, non-animal-based proteins that are found in soybeans. Presently, there is an increase in the addition of SP to food products due to its unique composition and nutritional content. A report by Anderson *et al.* showed that more than 11,000 food products contain SP [1]. Soybean is a legume and its plant is classified as an oilseed by the UN Food and Agriculture Organization (FAO). It varies in sizes and seed-coat colors (i.e., from black, brown, blue, yellow, green to uneven colors). Dry soybean contains 60% protein content, soy bean oil, and 35% carbohydrate (soluble and insoluble). The soybean oil contains phytosterols (Figure 5.1), namely brassicasterol, campesterol, sitosterol,

Corresponding author: blessingaderibigbe@gmail.com

Visakh P. M and Olga Nazarenko (eds.) Soy Protein-Based Blends, Composites and Nanocomposites, (103–134) 2017 © Scrivener Publishing LLC

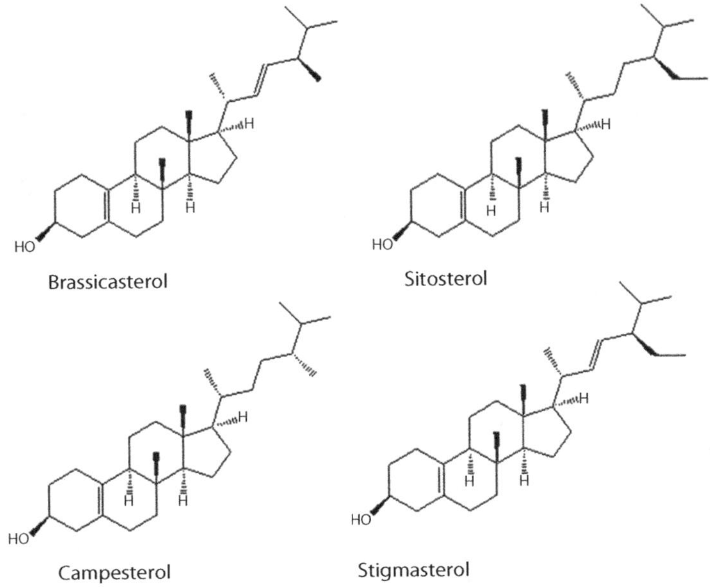

Figure 5.1 Structures of phytosterols found in SP.

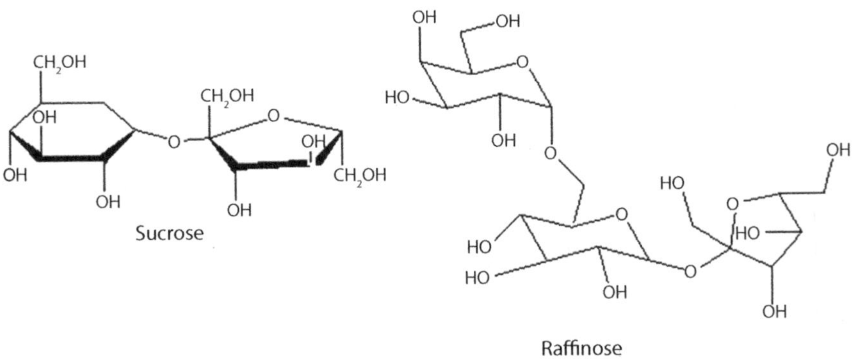

Figure 5.2 Structures of selected carbohydrates found in SP.

stigmasterol which can be converted into steroid hormones [2]. The oil has a good amount of omega-3 and omega-6. The soluble carbohydrates (Figure 5.2) found in soybean seed are sucrose (range 2.5–8.2%), raffinose (0.1–1.0%), and stachyose (1.4–4.1%) while the insoluble carbohydrates are cellulose, hemicellulose, and pectin [3]. Soybeans are referred to as a source of complete protein because it contains significant amounts of all the essential amino acids needed by the human body which the body is unable to produce [4, 5]. They are considered to be useful in the prevention of cancer and reduction of serum cholesterol levels [4]. Soybeans also contain a high amount of phytic acid that are antioxidant and can inhibit the growth of cancer cells, reduce blood sugar level and inflammation [6–8]. They are also a good source of fibre, iron, calcium, zinc, and vitamin B [9].

SP has several other health benefits such as it prevents bone loss and promotes a healthy heart in menopausal women and this is due to the present of estrogen called phytoestrogen [10]. Due to the great health benefit of SP, it is presently investigated in drug delivery devices, regenerative medicine, and tissue engineering. This chapter will be focused on the biomedical applications of SP.

5.2 The Forms of SP

The forms of SPs are texturized SP, SP concentrates, SP isolates, and SP flour and grits [11].

Texturized SP

Texturized SPs are defatted soy flours or concentrates which are mechanically processed by extruders and heating resulting in meat-like chewy textures. Extrusion cooking of SP flours or concentrates offers some benefits such as inactivation of trypsin inhibitors, control of bitter flavors, homogeneous bonding of ingredients that may include color, chemicals and other additives which can have effects on appearance and denaturing of the proteins [12, 13].

SP isolates

They are a refined form of SP with a protein content of about 90%. They are prepared by subjecting defatted soy flour to moist heat treatment so as to remove water insoluble carbohydrates followed by extraction with water and alkali at a pH of 7 to 8.5. The extract pH is adjusted to 4.5 resulting in precipitation of the proteins which are collected by centrifugation or filtration. The precipitated proteins are then washed, neutralized, and dried resulting in a neutral flavor [13]. They are used to improve the texture of meat products, to increase protein content, to enhance moisture retention, as an emulsifier, for scaffolds, wound-dressing materials, and drug delivery devices [13].

SP concentrates
They contain about 70% SP and it is prepared from defatted flakes or flour by washing with 60–80% alcohol resulting in the removal of sugars and other compounds that are soluble in alcohol. The concentrates are dried at an almost neutral pH. Acid leach is performed at pH 4.5 to remove the sugars and the wet protein concentrate is neutralized and dried. Moist heating is then applied so as to denature the proteins in the flakes or flour followed by a water wash to remove the sugars and other minor components [13]. They are used in a variety of food products such as in baked foods, breakfast cereals, and in some meat products. They are also used in meat and poultry products to improve the nutritional values. They exist in different forms such as flour, spray dried, and granules.

SP flour and grits
SP flour and grits contain 40–54% SP and they are prepared by grinding soybeans into a fine powder. They exist in different forms such as whole containing natural oils and defatted four in which the natural oil has been removed. SP flour is gluten-free [13]. SP flour has a 0.157 mm-sieve pore size where as grits are coarser.

5.2.1 Properties of SP

Although SP is widely used in food, it is a biomaterial that is used in scaffolds for tissue engineering and drug delivery devices. SP products can swell when they absorb water or can dissolve in water and this is an important functional property in drug delivery systems. This characteristic is applicable to plant-based protein products. The interaction of plant protein product with water can be influenced by various factors such as processing treatment during their manufacture and storage, hydrophilicity and hydrophobicity balance, surface amino acid composition, and environmental conditions including pH, ionic strength, and temperature [14,15]. Soybeans is the most cultivated plant in the world and as such, SP is readily abundant and cheap, it is biopolymer, renewable resource, biodegradable, biocompatible, and environmentally friendly [16–18]. It contains selected functionalities such as –OH, –SH, $-NH_2$ which are prone to chemical and physical modifications.

5.2.2 Application of Plant Protein in Biomedical Applications

Plant proteins are used in biomedical applications such as drug delivery systems and scaffolds for tissue regenerative medicine because protein makes up a major part of the body. Plant proteins are biocompatible and have the ability to maintain the functions of extracellular matrix when used as scaffolds. They can be tailored so as deliver bioactive agents to the body at selected pH because they contain functional groups that have different charges and this is dependent on the pH [19]. This is one unique characteristic that is lacking in synthetic polymers. Plant proteins are safer to use for medical applications because they do not pose the risk of disease transmission and potential immunogenicity as in the case of collagen, an animal-based protein which was reported to transmit mad cow disease. [20]. There are also reports of immunogenic effects of collagen on some patients [20]. Plant proteins are biodegradable, nonimmunogenic, and affordable, for example, corn, soybeans, and wheat. They have low molecular weight, high net negative charge, wide range of isoelectric points and they are considered suitable for delivery of positively charged bioactive agents [21]. They are hydrophilic because of the presence of polar amino acids and this attribute is responsible for their attraction to cells [21]. And because of the aforementioned characteristics of plant proteins, they are used to develop hydrogels, films, fibers, nanoparticles for tissue engineering, regenerative medicine, and drug delivery. However, plant proteins-based biomaterials have poor mechanical properties and poor hydrolytic stability and because of this, they are usually cross-linked or blended with synthetic polymers so as to improve its mechanical properties [21].

5.3 Wound-Dressing Materials

Wound healing is a process that is complicated, physiological and dependent on various factors such as enzymatic pathways, cellular and biochemical components. Selected biopolymers exhibit excellent structural and physicochemical properties which makes them suitable for preparation of wound-dressing materials. A wound-dressing material is usually prepared so as to be in direct contact with the wound. It is used to stop bleeding and to start clotting so that the wound can be healed, to absorb excess blood, plasma, or other fluids, to ease pain, to prevent infection of the wound, and to reduce psychological stress by covering the wound from the view of people and the patient [22]. Wound healing is dependent on a combination of inflammatory, vascular, connective tissue, and epithelial cells activities [22]. To reduce skin scar formation and speed-up the healing process, different wound dressings are presently in use. An ideal wound dressing exhibits selected characteristics such as [22,23] it provides mechanical and bacterial protection; it maintains a moist environment at the wound/

dressing interface; it allows gaseous and fluid exchange, it is nonadherent to the wound; nontoxic, nonantigenic, and nonallergic; it provides pain relief; highly absorbable; absorbs wound odor; easily sterilized; biodegradable, easy to use and affordable. Different types of wounds require different dressings or combination of dressings and there are different types of wound [22] dressing: dry dressings such as gauze and bandages, nonadhesive meshes, membranes and foils, foams, and tissue adhesives; moisture-keeping dressings such as pastes, creams, and ointments, nonpermeable or semipermeable membranes or foils, hydrocolloids, hydrogels, and combination products; bioactive dressings such as antimicrobial dressings, interactive dressings, single-component biologic dressings, and combination products [22].

5.3.1 Application of SPs in Wound Dressings

SP has advantage over other natural proteins and is used for wound dressings because of its low cost, non-animal origin, long storage time and stability. Specific amino acids in protein play an important role in wound healing, tissue regeneration, and building block components. These amino acids are arginine and glutamine. Arginine enhances efficient wound repair and immune function while glutamine is used by inflammatory cells in the wound for proliferation and as a source of energy [24–26]. SP contains arginine (7.5%) and glutamine (19%) (Figure 5.3) [27–28]. SP has a good affinity to human skin. Different forms of wound dressings are prepared from SP such as hydrogels, films, granules, paste, and nanofibres (Figure 5.4).

5.3.1.1 SP-Based Nanofibres for Wound Dressing

Nanofibres are generated by electrospinning. They exhibit unique properties such as increased surface area and reactivity, increased strength, ability to mimic extracellular matrix by acting as scaffolds that allow the cells to proliferate and high drug loading ability [29]. Due to their unique

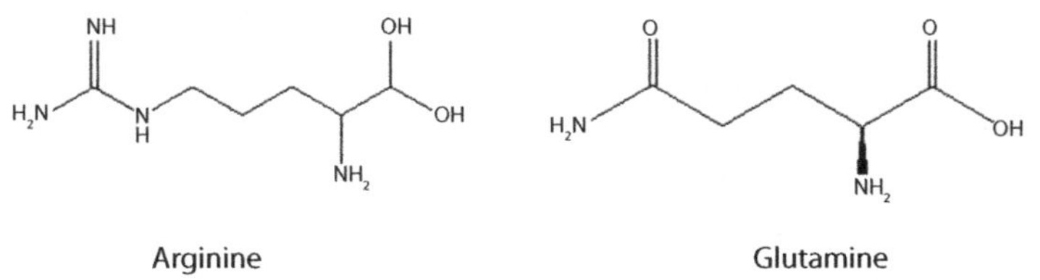

Figure 5.3 Structures of main compounds in SP responsible for enhanced wound healing.

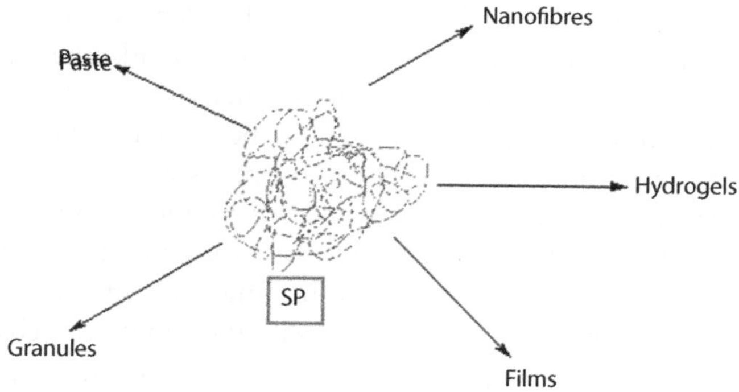

Figure 5.4 Different forms of wound dressings prepared from SP.

properties, they are used in wound dressing and tissue engineering. In wound dressing, their excellent pore densities and pores interconnectivity allow fluid to exude from the wound thereby avoiding the stagnating of fluid under wound dressing [29].

In a study by Har-el *et al.*, an electrospun SP-based tissue scaffolds were prepared for wound dressing and the wound-healing ability of the scaffolds was demonstrated using Yorkshire pigs [30]. The time and quality of wound-healing tissue were evaluated using histology (H&E and Masson's trichrome staining). Two weeks after a single application of SP-based scaffolds, the wounds showed robust signs of re-epithelialization and this was absent in the control wounds. After 4 weeks the wounds treated with SP scaffolds contained a stratified epithelial layer in the epidermis that looked essentially normal while the connective tissue in the dermis attained a cellular and organized appearance. In the untreated control wounds, the epidermis appeared immatured, while the dermis was still replete with numerous inflammatory/immune cells. Masson's trichrome staining further confirmed an increased presence of collagen in the dermis of the wounds treated with SP scaffolds at 4 weeks, while the control wounds were devoid of collagen. Sweat glands and hair follicles were observed after 4 weeks on the wounds treated with SP scaffolds and this was not observed in the untreated controls wounds. The scaffolds degraded rapidly during the wound-healing process and were fully adsorbed after 4 weeks. It was hypothesized that proteolytic degradation of the scaffolds might liberate some of the cryptic peptides contained in the SP which might enhance the wound-healing process by a mechanisms that is not understood yet [30, 31].

In another research report, a blend of SP isolate and poly(ethylene oxide) nanofiber web was prepared by electrospinning process [26]. The moisture vapor transmission rate result showed that the blended electrospun web provides suitable moist environment over the wound. The effect of the nanofiber web on wound healing was performed using female Wistar rats and excellent wound healing was visible by the growth of new epithelium without any significant adverse reaction. The nanofibre web exhibited good antibacterial activity against both *Staphylococcus aureus* and *Pseudomonas aeruginosa*. After 2 days, epithelial cells were formed on both the control wound and test wound but the rate of healing was different. The rate of healing was faster on the wound covered with the nanofibre web and this is attributed to the supply of glutamine from SP isolate for new cell proliferation and as a source of energy. On the fourth day, the test wound was fully covered and the thickness of the wound was reduced when compared to the control wound. On the eighth day, the control wound was fully covered with partial hair growth which indicated the growth of epithelial cells and the underlying dermal tissue was well vascularized [26].

5.3.1.2 SP-Based Films and Membranes for Wound Dressing

Protein-based films used for wound dressing are usually processed in water or under low moisture. SP-based films are prepared using several methods such as solvent casting, extrusion, solvent casting, spinning, and thermal compaction [32]. The properties of the films can be modified by the addition of plasticizers, cross-linkers, varying the amount of protein, and preparation conditions [32]. The films exhibit good water vapor transmission from the skin therefore avoiding stagnation of fluid under the wound dressing that can enhance infection and maceration.

Peles and Zilberman reported application of SP isolate-based films for wound dressing [32]. An antibiotic drug, gentamicin was incorporated into the films for local controlled release andfor protection against bacterial infection. The type of plasticizer, cross-linking agent and method of cross-linking were found to influence the tensile properties, the degree of water uptake, and the weight loss profile of the films. The water vapor transmission rate of the films was found to be in a range ~2300 g/m^2 per day. The release study kinetics showed that release profile of gentamicin from the films was sustained and maintained for at least 4 weeks. The release mechanism of gentamicin from the films was dominantly by diffusion. These films were found to be potentially useful wound dressing for burn and ulcer wounds [32].

In another research report by Peles *et al.*, homogeneous high-quality SP-based films encapsulated with gentamicin were tested for their effects on bacterial inhibition. The cytotoxicity and *in vitro* biocompatibility of these films were examined. The drug released from the films during an indirect cytotoxicity test was found to be safe and the biocompatibility test showed merging cell cultures in close proximity to the SPI films [33].

In some research reports, the physicochemical properties of potential SP-based wound dressing were studied. Silva *et al.* prepared blended membranes composed of chitosan and SP for wound healing and skin tissue-engineering scaffolding. The content of chitosan and SP was varied in the prepared membranes. Physicochemical properties of these membranes were studied and the findings suggested that they are potential wound dressing [34].

In a research study by Ramnath *et al.*, composite biomaterials prepared from SP and sago starch cross-linked with gluteraldehyde were prepared as temporary wound-dressing materials [35]. *In vivo* study was performed using 78 male albino Wister rats weighing 150–200 g. Full-thickness excision wounds were made on the back of male rats and the prepared wound dressing was applied on the wounds and changed periodically at an interval of 4 days. The wounds treated with the composite biomaterials prepared from SP and sago starch cross-linked with gluteraldehyde healed completely on twentieth day after wound creation, whereas control wounds showed complete healing only on the twenty-fifth day. There was also a significant increase in total collagen, hexosamine and uronic acid contents in the granulation tissues of the treated wounds. The prepared composite was found to be an excellent wound-dressing material due to its ability to absorb excess exudates, maintain a moist environment at the wound site, and enhance wound healing in the experimental animals [35].

There is also a report of membranes which was prepared from chitosan and SP isolate. Wound-healing ability of these membranes was studied on 20 male Sprague-Dawley rats that weighed between 230–280 g for a period of 2 weeks. The rat were divided into three groups (1) Group A: – wound directly covered with the prepared membranes; (2) Group B – wound directly covered with Epigard, a wound dressing that is used in clinical practice for short-term purpose; (3) Group C – no coverage of the wound. Wounds covered with the prepared membranes healed faster and re-epithelialization was observed a week after application of the membranes. This was followed by cornification of the outermost epidermal layer at the second week of dressing signifying repair of the wound tissue. These membranes possessed some other unique features such as ease of handling and final esthetic appearance [36].

5.3.1.3 *SP-Based Hydrogels for Wound Dressing*

Hydrogel dressings are available as gels, sheets, and gels preapplied to gauze. The gels are used for cavity wounds, and the sheet dressing and impregnated gauze can be applied to surface wounds [37]. Hydrogel dressings are useful for chronic wounds with slight exudation. These dressing have high water content and are able to keep granulation tissue and fresh epithelial tissue moist thereby protecting them from external mechanical stress and infection [38]. There is very little report on the SP-based hydrogels for wound dressing.

In a research report, biomimetic hydrogel for moist wound dressing application was prepared from poly(ethylene glycol) and SP with a water content of 96%. The relationship between the structural parameters of the network, its mechanical properties, and protein absorption *in vitro* was studied. Elastic moduli varied from 1 kPa to 17 kPa and this was dependent on the hydrogels composition. The Poisson's ratio was ≈0.18 and deformation at break ≈300% showed no dependence on this parameter. The reactions between polyethylene glycol and SP-created polymeric chains comprising of molecules of poly(ethylene glycol) and protein fragments between cross-links. The average molecular weight was three times higher than expected for a theoretical network [39].

Shigel *et al.* prepared and evaluated soft hydrogel system based on the poly(ethylene glycol)–protein conjugates as an occlusive wound-dressing material [40]. The hydrogel material contained 96% of liquid. It was formulated with phosphate-buffered saline and safe preservative to control bacterial load in the open wounds. The wound-dressing material was assessed in partial and full thickness wounds in pigs. The wounds treated with the hydrogels healed without any signs of inflammation, skin irritation, oedema, or erythema. Cellular composition of the re-epithelialized wounds was very similar to that of the normal skin, with a well-developed stratum corneum and epithelial layer. The hydrogel played an important role in providing pronounced hydration effect and maintaining a natural moist environment to the healing tissues. The hydrogels showed relatively low-protein-absorbing activity, absorbing predominantly low-molecular-weight molecules, including interleukin (IL)-1b, IL-6, transforming growth factor-b1 and products of haemoglobin degradation. The results obtained indicated that the wound dressing promoted fast re-epithelialization by creating favorable environment for keratinocytes proliferation and it also reduced scarring. The hydrogels were found to be biocompatible and inflammatory inert wound-dressing material [40].

Shevchenko and Santin reported the application of biomaterials that were formulated either as hydrogel or as a paste formulation. These formulations were assessed as wound dressings and dermal substitute *in vivo* using model pig skin. The wound dressing was changed every 3 day. Collagen deposition was observed and it is believed to be induced by the presence of SP. A unique pattern of early inflammatory response, eliciting neutrophils and controlling macrophage infiltration, followed by tissue cell colonization of the wound bed with a significant deposition of collagen fibres were observed. In large size, nonhealing wounds, wound dermis was best obtained with the paste formulation as hydrogels appeared to be too loose to ensure lasting scaffolding properties [41]. The general observations of the soybeans formulations in wound healing were that it exhibited a faster degradation time when compared toother collagen-based biomaterials; an inflammatory response that maintained neutrophil/macrophage balance in favor of the neutrophils, thus reducing the risks of infections, chronic inflammation that can result in scarring and the stimulation of collagen synthesis and angiogenesis promoting tissue repair in wounds not capable of spontaneous healing [41].

5.4 Potential Applications of SP in Regenerative Medicine and Tissue Engineering

Natural-based biomaterials have several advantages when compared to synthetic biomaterials. Some of these advantages are biocompatibility, biodegradability, and remodelling [42]. These advantages are useful in regenerative medicine and tissue engineering. Natural-based biomaterial can be further classified into: (1) protein-based biomaterials, for example,collagen, gelatin, silk, SP; (2) polysaccharide-based biomaterials, for example, cellulose, chitin/chitosan, glucose, and (3) decellularized tissue-derived biomaterials, for example, decellularized heart valves, blood vessels, liver [42]. Protein-based biomaterials are classified as animal- and plant-based biomaterials. Natural-based biomaterials are used to replace or restore structure and function of damaged tissues/organs because they can support cell adhesion, migration, proliferation, and differentiation [42]. When they are implanted into a defective tissue/organ, they enhance the attachment and migration of cells from the surrounding environment and thus induce extracellular matrix formation and promote tissue repair [42]. Animal-based proteins biomaterials suffer from some shortcomings such as toxicity [43], poor water stability [43], high potential of transmission of pathogens [44], and low cross-linking efficiencies and this has limited their use in tissue engineering and regenerative medicine. The use of SP or blends of SP with other polymers or macromolecules that enhanced cell proliferation has been reported. They have been reported to possess unique characteristics that make them potential materials that can beused in the repair and replacement of damaged human tissues and organs. Synthetic-based biomaterials exhibit some disadvantages such as their structure and composition is not similar to native tissues and organs, low biocompatibility, and their ability to induce tissue remodelling is very low [42].

In a research report by Silva *et al.*, SP-based membranes were prepared by cross-linking with glyoxal and tannic acid [45]. The membranes were subjected to *in vitro* degradation analysis for a period of 15 days. The membranes exhibit partial degradation and were found to be water resistant. They exhibited low rate of water transmission and this was attributed to the protein conformation, that is, globular protein. Adhesion and proliferation studies were performed for 7 days using SEM analysis. SEM images displayed a cluster of cells on the surface of the membrane. The membrane subjected to thermal treatment showed a film of cell formation on the surface of the membrane. Thermal treatment of the membranes reduced the cell adhesion ability [45]. However, the physical modification by thermal treatment had an outstanding effect on the membranes mechanical properties and *in vitro* biological behavior.

Chein *et al.* studied the biocompatibility of SP scaffolds fabricated by freeze-drying and three-dimensional printing [46]. The content of SP in the scaffolds was varied. It was assessed using a subcutaneous implant model in female BALB/c mice age 6–8 weeks. The size of the implants in the mice was measured using digital calliper 3 days after implantation and after every 3 days. The implanted scaffolds degraded over time with minimal scarring and no infections. One percentage containing SP scaffolds exhibited 100% degradation by day 15 while others degraded by day 18. No visible deformity as a result of the implant was seen at the implantation site. The skin and hair growth appeared normal and Masson's trichrome staining showed very little evidence of damage or fibrosis at the soy implant site. Scaffolds of higher SP content were still present after 56 days. H&E staining revealed that macrophage infiltration was hindered in the denser bioplotted soy scaffolds, causing slower degradation. Analysis of soy-specific antibodiesin mouse serum after implantation revealed levels of IgG1 that correlated with higher scaffold weight and protein density. Soy-specific IgE was not detected, signifying the absent of an allergic response to the soy implants. These results indicated that SP is a potential biocompatible implant for tissue regeneration. The scaffold porosity, SP density, and scaffold degradation rate significantly affected the acute and humoral immune response [46].

Chien and Shah prepared porous SP-based scaffolds [47]. SP was modified by heat treatment and enzyme cross-linking using transglutaminase in maltodextrin. The content of SP was varied in the prepared scaffolds. All the scaffold pore sizes were normally distributed in the range 10–125 μm. Enzyme treatment significantly lowered the moisture content. Increasing the amounts of appliedenzyme units, lowered the median pore size. Enzyme treatment did not affect the mechanical properties of the scaffolds. It increased the degradation time by a week. Human mesenchymal stem cells (hMSCs) were seeded and cultured *in vitro* on the scaffolds for a period of 2 weeks. The changes in scaffold degradation altered the growth and morphology of seeded hMSC. Cell proliferation was observed in scaffolds containing 3% SP isolate treated with 1 U of transglutaminase. The scaffolds enhanced cell attachment and cell viability for 2 week. In the 3% SP-containing scaffolds, cells proliferated with a threefold increase in DNA content was observed over the 2 weeks. The DNA content was constant over 2 weeks for the 5% SP-containing scaffolds. Three percentage SP-containing scaffolds showed that the cells either formed a cell sheet or integrated into the scaffold pores. These results demonstrated that controlling scaffold degradation rates is crucial for optimizing hMSC growth on SP scaffolds indicating their potential in tissue engineering applications [47].

In a research report by Silva *et al.*, a series of blended membranes were prepared from chitosan and SP isolate by solvent casting method [48]. Cross-linking was performed *in situ* with glutaraldehyde solutions in the range $5 \times 10^{-3} - 0.1$ M. The obtained results showed that the incorporation of SP isolate and chitosan to the network by cross-linking, promoted a slight decrease in water absorption and decreasedthe degree of degradability of the membranes. A biocompatibility study was performed on the membranes using L929 cells and the results obtained indicated that the membranes support cell growth [48].

Xu *et al.* reported nonwoven nanofibre membranes that were prepared from SP isolate and polyethylene oxide by dissolving them in 1,1,1,3,3,3-hexafluoro-2-propanol [49]. They were prepared from the solution by electrospinning. Polyethylene oxide was used as a co-spinning polymer so as to improve the spinnability of SP isolate. The ratio of SP to polyethylene oxidewas varied. The diameter of most of the nanofibres was in the range of 200–300 nm. All the nanofibre membranes showed superhydrophilicity and were found to be useful in biomedical applications [49].

Xu *et al.* reported the preparation of water-stable electrospun SP-based scaffolds [43]. The scaffolds had large volume and ultrafine fibres oriented randomly and evenly in three dimensions. They were used to simulate native extracellular matrices of soft tissues. The disulphide cross-links in SP were cleaved to facilitate its dissolution in an aqueous solvent system. The solution obtained was then electrospun into bulky scaffold. The three-dimensional ultrafine fibrous SP scaffolds demonstrated long-term water stability and maintained their fibrous structures after incubation in PBS for up to 4 weeks. *In vitro* study was performed on two-dimensional scaffolds and the three-dimensional ultrafine fibres scaffolds. Cells could not be observed 60 μm below the surface of the SP two-dimensional scaffold. The distribution of cells in each plane was uniform. The tight packing of fibres in the horizontal direction and lack of space in the thickness direction are believed to may have contributed to the low penetration of cells in the two-dimensional structures, resulting in a neotissue of this scaffolds. This shortcoming indicated that they cannot be used in large volume reconstruction in soft tissue regeneration. Three-dimensional SP fibrous scaffolds exhibited higher proliferation rates of the adipose-derived mesenchymal stem cells. The amount of cells attached on three-dimensional scaffolds was 163% and 210% times more than the two-dimensional scaffolds. After 2 weeks of cell culture, the proliferation of cells on SP three-dimensional fibrous scaffolds was 227% and 114% higher than the two-dimensional scaffolds. The tight packing of fibres in the two-dimensional scaffolds hindered the penetration of cells vertically, whereas the multiple pores and high porosity of the three-dimensional scaffolds enhanced the migration and penetration of cells into the interior of the structures. *In vitro* study showed that the three-dimensional SP scaffolds supported uniform distribution and adipogenic differentiation of adipose-derived mesenchymal stem cells when compared to two-dimensional SP scaffolds [43].

In another research report, the parameters for electrospinning fibrous scaffolds from SP isolate by the addition of poly(ethylene oxide) dissolved in 1,1,3,3,3-hexafluoro-2-propanal were investigated. Their physicochemical properties were studied and they were found to exhibit mechanical properties that are similar to human skin. They promoted proliferation of cultured primary human dermal fibroplast alone and in organic co-culture with primary human epidermal keratinocytes. Similar gene expression profiles of fibroplasts cultured on SP isolate substrate were seen at time 2 h and 24 h [50].

Santos also reported the preparation of hybrid membranes from chitosan and SP isolate [51]. They were prepared by a combination of a sol–gel method and solvent casting. The effect of these membranes on the activation of human polymorphonuclear neutrophils (PMNs) was studied. The effect of the biomaterials on the activation of PMNs was assessed by the quantification of lysozyme and reactive oxygen species. *In vitro* studies revealed that the membranes exhibited low stimulation of human PMNs for the release of lysozyme and for the production of ROS found in the respiratory burst. The low *in vitro* stimulation of the PMNs induced by these membranes indicated that they can promote normal wound-healing process and restoration of tissue function when implanted *in vivo* [51].

Shevchenko and Satin demonstrated the healing properties of soybean-based hydrogels and pastes in two pre-clinical *in vivo* models based on healing (full-thickness punch biopsy model) and nonhealing (full-thickness chamber model) [41]. Two paste formulations were prepared by mixing granules with hydrogels without or with genipin cross-linking: (1) Soybean-based granules (40%) and soybean-based hydrogel (60%); (2) Soybean-based paste cross-linked with 1% (w/v) genipin. All the wounds treated with soy-containing biomaterials showed signs of inflammation after 8 days of implantation. This acute inflammatory reaction toward the soybean-based hydrogel was resolved by day 15 and the material was completely reabsorbed and tissue cells populated the wound site depositing new extracellular matrix. The paste formulations were stable and exhibited tissue scaffolding properties over a longer period of time and were still present after 25 days of implantation. The hydrogels exhibited fast degradation thereby creating space among the more stable granules that gradually encouraged tissue infiltration, first in form of acute inflammatory response exhibiting a prevalence of neutrophils to that of macrophages, later inducing the thorough invasion of new dermis where collagen and vessels gradually progressed from the bottom of the wound toward its surface [41].

Silva *et al.* reported soy- and casein-based membranes for biomedical applications [45]. The membranes were subjected to cross-linking with glyoxal and tannic acid followed by thermal treatment. The cytotoxicity of both soy- and casein-based protein biomaterials was evaluated and it correlated with the materials degradation behavior. Soy- and casein-based protein used did not exert any cytotoxic effect over cells. However, they enhanced cell proliferation [45].

Electrospun fibers were prepared from SP isolate having an average fiber diameter in the nanometer range for tissue engineering applications. Fiber formation was enhanced by the addition of poly(ethylene oxide). The SP isolate/poly(ethylene oxide) mats were cross-linked using carbodiimide to increase its robustness. SP isolate/poly(ethylene oxide) fiber diameters ranged between 50 nm and 270 nm depending on electrospinning and solution parameters. The Young's modulus for 7% SP isolate/3% poly(ethylene oxide) and 12% SP isolate/3% poly(ethylene oxide) electrospun scaffolds were 75 kPa and 252 kPa, respectively. *In vitro* analysis showed that hMSC adhesion and proliferation on the SP isolate/poly(ethylene oxide) fibers, indicating their potential applications in tissue engineering [52].

In another research, SP slurry was successfully printed using the three-dimensional bio-plotter to form scaffolds. The structural integrity of the scaffolds during printing was evaluated. The optimal mass flow rate for printing soy slurry at 27 °C was 0.0072±0.0002 g/s. The addition of dithiothreitol to soy slurries demonstrated the importance of disulfide bonds in forming solid structures upon printing. Resulting bio-plotted SP scaffolds were cured using 95% ethanol and post-treated using dehydrothermal treatment (DHT), a combination of freeze-drying and DHT and chemical cross-linking using 1-ethyl-3-(3 dimethylaminopropyl) carbodiimide. Seeding efficiency of hMSCs was highest for nontreated and thermally treated scaffolds and all the scaffolds supported hMSCs viability over time [53].

Chien *et al.* reported SP injectable hydrogels that were prepared without chemical modifiers or cross-linkers [54]. *In vivo* viability and growth of seeded L929 mouse fibroblasts demonstrated that the hydrogels were biocompatible *in vitro* for 1 week. Soy hydrogels were injected into the subcutaneous pocket of mice and histological staining showed minimal fibrous capsule formation up to 20 days. It was found to be a potential biomaterial for tissue engineering and drug delivery applications [54].

5.5 Application of SP Product for Regeneration of Bone

Soy protein is being investigated for their ability to repair bone defects. Materials used for regeneration of bones are required to be easy to handle, biodegradable, noncytotoxic, nonimmunogenic and must be able to induce bone regeneration. Soybean is an attractive source of biomaterial for bone repair because it contains bioactive phytoestrogens that can induce differentiation of osteoblasts (i.e., bone-forming cells) [55]. It also exhibits some of the characteristics required for materials used in bone regeneration. Soybean-based biomaterials used in bone regeneration are prepared by thermosetting of soybeans flour which are then processed into different forms [55] depending on the applications such as bone fillers, hydrogels, composites, granules, membranes, and composite foams. The soybean-based bone fillers are ductile, adapt easily to the implantation site and are potential biomaterials for applications in orthopaedic, maxillofacial and periodontal surgeries [55]. They absorb water and exhibit rubbery consistency, biocompatible, biodegradable, and cost competitive [56]. Currently, materials such as calcium phosphate cement, bone marrow replacement and auto grafting, allografting cancellous bone, and vascularized bone grafts are used for regeneration of bone. Their use in clinical application is limited by their relatively slow rate of resorption, scarce tissue donor, risks of transmittable diseases, infections, inconsistent clinical performances, expensive and some require major operation with sophisticated infrastructure. These limitations have prompted the development of alternative biomaterials [57–63].

In a research study by Vac *et al.*, SP thermoplastics were prepared by varying the amount of glycerol, ceramic reinforcement, and coupling agents [64]. The tensile and the dynamic-mechanical behavior of injection-molded samples of the prepared SP thermoplastics were evaluated. The incorporation of glycerol into a soy-based matrix resulted in its plasticization which was confirmed by the drop in stiffness, that is, storage and elastic modulus above 20 °C and a decrease in the protein glass transition temperature. Differential scanning calorimetric thermographs showed the occurrence of conformational changes in the SP during processing. The prepared thermoplastics exhibited thermal stability up to 100 °C which was confirmed by thermogravimetric analysis. The reinforcement of the SP matrix with ceramic filler (tricalcium phosphate) was found to be effective for amounts above 10% w/w. Amino-coupling agent used resulted in a plasticizing effect that was displayed in the mechanical and dynamic-mechanical properties of the thermoplatics. The properties exhibited by these thermoplastics further confirmed that SP is a potential biomaterial for orthopedic applications [64].

Satin *et al.* prepared biodegradable biomaterials by thermosetting of defatted soybean curd [65]. These biomaterials were found to be useful for preparation of films, porous scaffolds, and granules for different surgical needs. The biomaterial was incubated in physiological solutions and the water uptake of 80%, elongation at break of 0.9 mm/mm, and 25% (w/w) degradation in 7 days. The biomaterial granules reduced the activity of the monocytes/macrophages and of the osteoclasts and induced osteoblast differentiation *in vitro*. These findings further demonstrated the bone regeneration potential of soybeans-based biomaterial [65].

Natural, biodegradable soybeans-based hydrogels were prepared by the thermosetting of defatted soy curd. These hydrogels regulated the activity of both tissue and inflammatory cells. The content of the different soy components such as the isoflavones were tuned varying the solvent system during the extraction procedure. The variations in the material cross-linking provided either loose hydrogels or a bio-glue. They were able to stimulate the synthesis of collagen by fibroblasts and the formation of mineralized bone noduli by osteoblasts [66].

In another research report, the response of osteoblasts to injectable foamed bone cement based on a composite formulation containing foaming agents such as soybean and gelatine was studied. A self-hardening soy/gelatine/hydroxyapatite composite foam was prepared and it was able to retain porosity upon injection. The foamed paste produced a calcium-deficient hydroxyapatite scaffold after setting. It exhibited good injectability and cohesion as well as interconnected porosity after injection. The intrinsic bioactivity of soybean and gelatine promoted osteoblast adhesion and growth. These findings indicated that soybean can be used in combination with calcium phosphate cementsfor bone regeneration [67].

Bone auto graft which is the current practise of bone filling is limited by material availability, risks of infection, and donor site morbidity. The presently used materials as bone fillers do not show all the necessary characteristics such as osteoconductivity, controlled biodegradation, and ease of adaptation to the implantation site. Merolli *et al.* reported a class of biodegradable material based on soybeans [68]. These materials showed good mechanical properties and an intrinsic bioactivity on inflammatory and tissue cells *in vitro*. *In vivo* study of these biomaterials was performed in repairing a surgical lesion and comparing it with a sham-operated contralateral lesion of critical size (nonhealing model). Twenty-six operations were performed on white rabbits. Implantation of the soybeans biomaterial over a period of 8 weeks produced bone repair with features distinct from those obtained by healing in a nontreated defect. New and progressively maturing trabeculae appeared in the animal group where soybeans biomaterial granules were implanted, whereas the sham operation produced only a rim of pseudo-cortical bone still featuring a large defect. The trabeculae formed in the presence of soybeans biomaterial granules had features typical of reticular bone. These indicated that soybeans are potential bone filler in bone regeneration [68].

Givaresi *et al.* compared the bone regeneration potential of a soybean-based bone filler with a commercial 50:50 poly(d,l-lactide–glycolide)-based bone graft (Fisiograft® gel) [69]. They were implanted into a critical size defect (6 mm diameter, 10 mm length) in rabbit distal femurs. The histomorphometric and micro hardness analysis of femoral condyles was monitored for 4, 8, 16, and 24 weeks after surgery. No significant difference was found in the percentage of both bone repair and bone in-growth in the external, medium, and inner defect areas. Soybean-based filler-treated defects showed a significantly high outer bone formation and microhardness results at 24 weeks when compared to Fisiograft® gel ($P < 0.05$). Soybean-based biomaterial was found to promote bone repair through a mechanism of action that involved both the scaffolding role of the biomaterial for osteoblasts and the induction of their differentiation [69].

Chitosan and soybean protein isolate blended membranes were prepared by solvent casting. These membranes exhibited a biphasic structure that originates *in situ* porous formation, through a two-step degradation mechanism. These membranes were found to be suitable for barrier membrane in guided bone regeneration. The effect of β-radiation over the properties of these membranes was evaluated at three different doses (25 kGy, 50 kGy, and 100 kGy) and the surface chemical changes and bulk properties changes were assessed. No substantial changes were detected on the studied properties and the only observation was that the surface energy was found to be slightly increased for higher applied doses [70]. β-radiation was found to be a suitable sterilization method of the membranes.

Composite hydrogels based on the combination of natural polymers such as alginate, SP isolate, bioactive glass particles were reported by Silva *et al.* [71]. A sonochemical approach was used for the formation of rapid homogeneous composite hydrogels. Biomineralization process in simulated body fluid was observed over time and the results demonstrated that the composite materials have the ability to form a surface apatite layer after 7 days. This indicated that the material is suitable for bone regeneration applications.

5.6 Application of SP in Drug Delivery Systems

Protein-based biomaterials are potential matrices for controlled drug delivery systems. They have been studied extensively because of their biodegradability, biocompatibility, and availability. Their protein networks have the ability to interact with a range of bioactive compounds via their functional groups. These functional groups are useful for reversible binding of bioactive molecules thereby protecting the bioactive moleculesfrom enzymatic attack until they are released at the target site in the body [19, 72]. Hydrolysis of SP by digestive enzymes generates bioactive peptides that may exert a number of physiological effects *in vivo* on the gastrointestinal, cardiovascular, endocrine, immune and nervous systems [73, 74]. A wide variety of protein-based drug delivery formulations such as films, hydrogels, biocomposites, microparticles, and nanoparticles have been reported (Figure 5.5).

Chen *et al.* reported the controlled release mechanism of SP isolate-based films in the absence or presence of digestive enzymes at gastric or intestinal pH [75]. SP isolate denatured by heating was cooled before adding either methylene blue or rifampicin, followed by addition of formaldehyde as a cross-linking agent to modify film solubility, puncture force and resistance

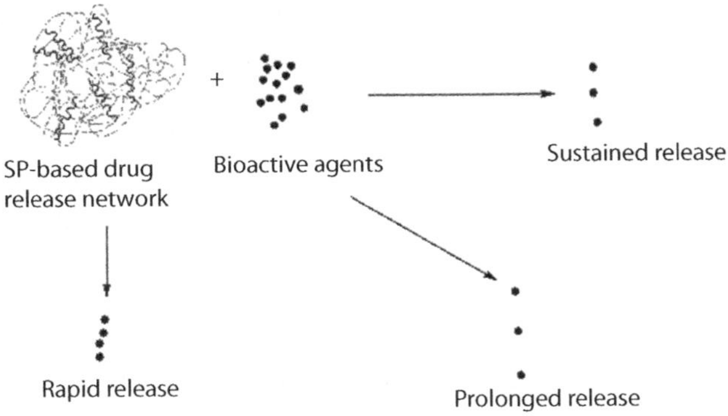

Figure 5.5 Different forms of SP-based drug delivery formulations.

under gastrointestinal conditions. The effect of formaldehyde concentration on film properties and on the release of the compounds under gastrointestinal conditions was studied. The film networks displayed excellent compound binding capacity especially for hydrophobic molecules and the film bulk erosion in the presence of digestive enzymes followed first-order kinetics. Methylene blue or rifampicin release followed variable kinetics depending on compound solubility during a 1–2 h initial phase, followed by zero-order release. Cross-linking density was found to be an effective means of regulating the erosion and the release rate of the films.

Vaz *et al.* reported SP drug delivery matrix systems prepared by melt-processing techniques, namely extrusion and injection moulding [76]. The soy matrix systems were encapsulated with theophylline drug by extrusion and cross-linking with glyoxal. It was performed at two different pH values: 4 and 7, followed by injection moulding into a desired shape. The matrices were reinforced with bioactive filler, hydroxylapatite. The drug release from the matrices was affected by swelling, drug diffusion, formulation composition, and polymer dissolution. The formulation composition of the matrices was varied and it influenced the drug release rates. At pH 7.4, increasing the degree of cross-linking of the polymer matrix or decreasing the net charge resulted in decreased rate of drug release. The incorporation of ceramic filler accelerated the rate of drug release from the matrices and this was attributed to lack of interaction of the ceramic filler thereby, resulting in rapid drug diffusion through the open polymer ceramic interfacial regions. At low pH, no significant difference was observed in the rate of drug release [76].

Vaz *et al.* developed SP double-layer controlled-release device based on a nonconventional approach [77]. Premixes of the SP isolate, fillers and the drug of choice, theophylline were prepared prior to processing. The premixes were converted into plastic materials by addition of extrusion liquids such as the plasticizers (glycerol and water) and a cross-linker (glyoxal). The extrusion process was performed at a barrel temperatures ranging from 70 °C to 80 °C. After the extrusion process, the extrudates were cooled to room temperature and cut into pellets. The properties of the extrudates ranged from a thermoplastic to a thermoset behavior, depending on the amount of cross-linking achieved during extrusion. The drug release behavior of the prepared double-layer device was compared with a single-layer device. The single-layer device exhibited initial burst effect and this was not observed in the double-layer device. The double layer was useful in reducing the surface available for drug release and in limiting the solvent penetration rate. Thus, the bursteffect observed in the single-layer device was controlled and the area available for drug release was maintained at a relatively constant level in the double-layer device [77].

Reddy *et al.* demonstrated the potential of SP isolate films as a drug release system for naturally occurring antiproliferative agent [78]. The films were prepared by casting method and the percentage of the resorcinol was varied between 10% and 30%. After the preparation of the films, the dried SP isolate films were peeled-off from the mould, sandwiched between two aluminium sheets and cured by hot pressing at 100 °C for 20 min under a 2 MPa pressure. The drug loading efficiencies of the films were found to decrease with increase in the cross-linking of SP isolate with resorcinol. The drug release study was performed over a period of 24 h. The release rate of curcumin an antiproliferative agent from the films decreased with an increase in the degree of cross-linking of the films. The release mechanism suggested that SP isolate films can improve the overall therapeutic effectiveness of curcumin by improving its bioavailability and thus decreasing the rapid metabolism and systemic elimination which is a pharmaceutical shortcoming of curcumin [78].

Nayak *et al.* reported the preparation and application of SP isolate/modified montmorillonites-based nanocomposites films [79]. They were prepared by blending SP isolate with different amount of modified montmorillonites by melt extrusion technique. The delivery mechanism of ofloxacin from the nanocomposite films was investigated at pH 3.4 and 7.4. Ofloxacin was released more rapidly at pH 7.4 than at pH 3.4. This is because electrostatic interaction of composites was easily broken at pH 7.4 than at pH 3.4 resulting in rapid release at pH 7.4 than pH 3.4. This finding suggested that the drugs in the blend are suitable for the large intestine, colon and rectal mucosa where there are different emptying times. The release profiles of drug from composites films loaded with different amounts of the drug was investigated. The release profile suggested that the formulation containing the highest amount of drug (50%) displayed a fast and higher release rate than those formulations containing a small amount of drug. The release rate was slower in films loaded with less percentage of drug and this is attributed to the presence of more free spaces through which a lesser number of drug molecules could transport [79].

Saamal *et al.* reported the development of nanocomposites by cross-linking SP with furfural and blending with organoclay (Cloisite 30B) [80]. It was used to demonstrate the controlled release studies of cefadroxil, an antibiotic drug. The *in vitro* cumulative drug release was performed at acidic and alkaline pH. An enhanced release profile was observed at alkaline pH which is because the hydrogen bond between the two moieties breaks down resulting in the release of the drug. An increased drug loading from 15% to 25% influenced the swelling of the nanocomposites by enhancing swelling and thereby resulting in the diffusion of the drug out of the polymer matrix [80].

Pradhan *et al.* also reported nanocomposites prepared by blending SP isolate with different percentage of cloisite 30 b by melt extrusion technique [81]. It was used to study the release mechanism of curcumin from the nanocomposites. The release of curcumin from the nanocomposites was faster at pH 7.4 than at pH 1.2 [81].

Electrospun nanofibers was fabricated from SP isolate/poly(ethylene oxide) blend and poly(lacticacid) [82]. Release studies of allylisothiocyanate, an antimicrobial bioactive agent was performed under dry conditions and the release of allylisothiocyanate was negligible. The release of allylisothiocyanate increased dramatically under relative humidcondition. This effect was attributed to the plasticizing effect of water on SP isolate which increased the molecular motion in thematrix thereby enhancing the release of allylisothiocyanate [82].

He *et al.*, prepared food protein-based nanoemulsions for drug delivery from whey protein isolate, soybean protein isolate and β-lactoglobulin [83]. Soybean protein isolate-based emulsion was prepared by dispersing soybean protein isolate powder into deionized water with stirring at 25 °C. The solution was adjusted to pH 7.0 using sodium hydroxide solution. The nonpolar and disulfide bonds in the protein interior were denatured by heating at 105 °C resulting in an increased emulsifying capacity of the proteins. The denatured protein solution was cooled to 25 °C. The nanoemulsion was encapsulated with fenofibrate, a highly lipophilic drug that is used to lower lipid levels. *In vitro* cytotoxicity analysis was performed on the nanoemulsion using Caco-2 cells and it exhibited good biocompatibility than the traditional emulsifier. *In vivo* absorption was performed on rats and the bioavailability of protein-stabilized nanoemulsion was increased significantly compared to the control. The oil/water nanoemulsion system exhibited good biocompatibility allowing for better and more rapid absorption of drugs [83].

In a research report by Liu, Cui, and Liao, pH and temperature responsive interpenetrating polymer network (IPN) hydrogels were prepared from SP and poly(*N*-isopropylacrylamide-co-sodium acrylate [84]. The hydrogels had porous honeycomb structures, good miscibility, thermal stability and good pH- and temperature-responsive. Varying the SP or cross-linker content was used to control the swelling behavior of the hydrogels. The hydrogels displayed the fastest deswelling rate in pH 1.2 and at 45 °C. Release study was performed and bovine serum albumin release from the hydrogels was pH and temperature dependent [84].

Yin *et al.* developed long-term stable oil in water emulsion from SP and soy polysaccharide [85]. A high pressure homogenization produced the protein/polysaccharide complex emulsion havinga droplet size about 250 nm. Heat treatment of the emulsion resulted in the protein denaturation and development of irreversible oil-water interfacial films composed of SP/soy polysaccharide complexes. It was found to be suitable delivery system for lipophilic bioactive agents [85].

In a research study by Chien *et al.*, SP hydrogels were developed by varying the weight percentages of water (15 wt%, 18 wt%, and 20 wt%) [55]. Chemical modifiers or cross-linkers were not used to prepare the hydrogels. This method was useful for developing hydrogels for direct injection *in vivo*. The concentration of SP was varied and it influenced the rheological, swelling, mechanical properties and the release of the model drug, fluorescein from the hydrogels *in vitro*. Higher weight percent of soy enhanced the robustness of the hydrogels and it resulted in the extended release rate of fluorescein for a week [55].

Maltais, Remondetto, and Subirade prepared SP filamentous hydrogels and tablets-based drug delivery systems [86]. The swelling and drug ability of these hydrogels were performed in simulated gastrointestinal conditions. Riboflavin was selected as a drug of choice and its release from these drug delivery systems were investigated under simulated gastrointestinal conditions in the presence or absence of digestive proteases. At intestinal pH the hydrogels shrinked after 1 h. Riboflavin release was faster at pH 7.5 than at pH 1.2 for both drug delivery systems. Swelling was reported to be the principal mechanism of riboflavin release from tablets at pH 7.5. At pH 1.2, drug-polymer interactions slowed drug release. In the presence of pepsin at pH 1.2, both systems exhibited slow zero-order release of riboflavin for 6 h, whereas at pH 7.5, both systems digested completely in the presence of pancreatin [86].

Fast temperature-responsive interpenetrating polymer network hydrogels based on SP and poly(*N*-isopropylacrylamide) (PNIPAAm) were also reported [87]. They were prepared using the sodium bicarbonate solutions as the reaction medium. The hydrogels have high porous structures and a fast release rate. The release mechanism of bovine serum albumin was anomalous transport [87].

Liu and Cui also reported temperature-sensitive interpenetrating polymer network hydrogels prepared from SP and poly(*N*-isopropylacrylamide) [88]. The hydrogels exhibited good miscibility, thermal stability and were temperature sensitive. Changing the content of SP or cross-linker was found to control the swelling behavior, water retention and network structure of the IPN hydrogels suggesting their potential application in drug delivery systems [88].

5.7 Conclusion

Soy protein is useful in wound dressing, drug delivery devices and has potential in regenerative medicine such as in bone regeneration. SP is a biomaterial that is biodegradable, nonimmunogenic, affordable, it has low molecular weight, high net negative charge, wide range of isoelectric points suitable for delivery of positively charged bioactive agents. It is hydrophilic because of the presence of polar amino acids which is responsible for its attraction to cells. However, it has poor mechanical properties and poor hydrolytic stability and because of this, it is usually cross-linked or blended with synthetic polymers so as to improve its mechanical properties. In its application in wound dressing, it contain specific amino acids, that is, arginine and glutamine which play an important role in wound healing, tissue regeneration and building block components. Arginine enhances efficient wound repair and immune function while glutamine is used by inflammatory cells in the wound for proliferation and as a source of energy. SP contains 7.5% arginine 19% and glutamine. The reported different forms of wound dressings prepared from SP are hydrogels, films, granules, paste and nanofibres. Nanofiber web of SP used as wound dressing displayed excellent wound healing and good antibacterial activity against both *S. aureus* and *P. aeruginosa* and increased the rate of wound healing. SP-based films and membranes prepared for wound dressing were reported to absorb excess exudates from the wounds, maintain a moist environment at the wound site and enhance wound healing in the experimental animals. Hydrogel dressings were reported to be useful for chronic wounds with slight exudation. These dressing were able to keep granulation tissue and fresh epithelial tissue moist thereby protecting them from external mechanical stress andinfection. However, there is very little report on the SP-based hydrogels for wound dressing. The results from the research reports confirmed the potential of SP in wound dressing.

There is very little research report on SP or blends of SP with other polymers or macromolecules for enhanced cell proliferation. SP support cell adhesion, migration, proliferation, and differentiation. In the selected research reports, SP scaffolds enhanced cell adhesion, proliferation, and differentiation indicating their potential in regenerative medicine. SP is also an attractive biomaterial for bone repair because it contains bioactive phytoestrogens that can induce differentiation of osteoblasts. Different forms of SP-based biomaterial have been developed for bone repairs such as: bone fillers, hydrogels, composites, granules, membranes, and composite foams. The bone fillers are ductile, adapt easily to the implantation site and are potential biomaterials for applications in orthopaedic, maxillofacial, and periodontal surgeries. They absorb water and exhibit rubbery consistency, biocompatible, biodegradable, and cost competitive. They were able to stimulate the synthesis of collagen by fibroblasts and the formation of mineralized bone noduli by osteoblasts which indicated their potential application in bone regeneration. Their protein networks of SP suggest its ability to interact with a range of bioactive compounds via its functional groups. These functional groups are useful for reversible binding of bioactive molecules thereby protecting the bioactive molecules from enzymatic attack until they are released at the target site in the body. Hydrolysis of SP by digestive enzymes generates bioactive peptides that may exert physiological effects *in vivo* on the gastrointestinal, cardiovascular, endocrine, immune and nervous systems. A wide variety of protein-based drug delivery formulations have been reported such as films, hydrogels, biocomposites, microparticles, and nanoparticles. They were found to be pH and temperature dependent, the rate of drug release from these devices was influenced by the degree of cross-linking of SP. Although SP has great potential in wound dressing and regenerative medicine, a better understanding of the mechanism of wound-healing process of SP-based wound dressing is needed. There is also a pressing need to develop these SP-based biomaterials to the stage of clinical trials.

Acknowledgement

The financial assistance of National Research Foundation and Medical Research Council (Self-Initiated Research), South Africa towards this research is hereby acknowledged. The views and opinions expressed in this manuscript are those of the authors and not of MRC or NRF.

References

1. Anderson, J.W., Johnstone, B.M., Cook-Newell, M.E., Meta-analysis of the effects of soy protein intake on serum lipids. *New Engl. J. Med.*, **333**, 276–282, 1995.
2. Wang, F.Q., Yao, K., Wei, D.-Z., From soybean phytosterols to steroid hormones, Chapter 11, Soybean and health, book edited by Hany El-Shemy, ISBN 978-953-307-535-8, Published: September 12, 2011. http://www.intechopen.com/books/soybean-and-health/from-soybean-phytosterols-to-steroid-hormones
3. Obendorf, R.L., Kosina, S.M., Chapter 13, Soluble carbohydrates in soybean, agricultural and biological sciences » *Soybean - biochemistry, chemistry and physiology*, book edited by Tzi-Bun Ng, ISBN 978-953-307-219-7, Published: April 26, 2011. http://www.intechopen.com/books/soybean-biochemistry-chemistry-and-physiology/soluble-carbohydrates-in-soybean
4. Henkel, J. Soy: health claims for soy protein, question about other components. FDA Consumer (Food and Drug Administration), 34, 18–20, 2000.
5. Dudek, S.G., Nutrition essentials for nursing practice, 4th edn., Philadelphia: Lippincott, 2001.
6. Vucenik, I., Shamsuddin, A.M., Cancer inhibition by inositol hexaphosphate (IP6) and inositol: from laboratory to clinic. *J. Nutr.*, 133, 3778S–3784S, 2003.
7. Yoon, J.H., Thompson, L.U., Jenkins, D.J.A., The effect of phytic acid on *in vitro* rate of starch digestibility and blood glucose response. *Am. J. Clin. Nutr.*, 38, 835–842, 1983.
8. Sudheer, K.M., Sridhar, R.B., Kiran, B.S., Bhilegaonkar, P.M., Shirwaikar, A., Unnikrishnan, M.K., Anti-inflammatory and anti-ulcer activities of phytic acid in rats. *Ind. J. Exper. Biol.*, 42, 179–185, 2004.
9. Lindsay, S.H., Claywell, L.G., Considering soy: its estrogenic effects may protect women. *AWHONN Life lines*, 2, 41–44, 1998.
10. Wardlaw, G.M., *Contemporary nutrition*, 4th edn., Boston: McGraw Hill, 2000.
11. Wolf, W.J., Soybean proteins: their functional, chemical, and physical properties. *J. Agric. Food Chem.*, 18, 969–76, 1970.
12. Hayakawa, I., Nayashi, N., Urushima, T., Fujio, F., Texturization of whole soybeans in a twin screw extruder. *J. Faculty Agron Kyushu Univ.*, 33, 213–220, 1988.
13. Singh, P., Kumar, R., Sabapathy, S.N., Bawa, A.S., Functional and edible uses of soy protein products. *Compr. Rev. Food Sci. Food Saf.*, 7, 15–26, 2008.

14. Moure, A., Sineiro, J., Domínguez, J.C., Parajó, J.C., Functionality of oilseed protein products: a review. *Food Res. Int.*, 39, 945–963, 2006.
15. Lee, K.H., Ryu, H.S., Rhee, K.C., Protein solubility characteristics of commercial soy protein products. *J. Am. Oil Chem. Soc.*, 80, 85–90, 2003.
16. Cho, S.Y., Rhee, C., Mechanical properties and water vapor permeability of edible films made from fractionated soy proteins with ultrafiltration, *Lebensmittel-Wissenschaft und Technologie*, 37, 833–839, 2004.
17. Zhang, J., Mungara, P., Jane, J., Mechanical and thermal properties of extruded soy protein sheets, *Polymer*, 42, 2569–2578, 2001.
18. Malafaya, P.B., Silva, G.A., Reis, R.L., Natural–origin polymers as carriers and scaffolds for biomolecules and cell delivery in tissue engineering applications. *Adv. Drug Deliv. Rev.*, 59, 207–233, 2007.
19. Chen, L., Remondetto, G.E., Subirade, M., Food protein-based materials as nutraceutical delivery systems. *Trends Food Sci. Tech.*, 17, 272–283, 2006.
20. Peng, Y.Y., *et al.*, Evaluation of the immunogenicity and cell compatibility of avian collagen for biomedical applications. *J. Biomed. Mater. Res.*, 93A, 1235–1244, 2010.
21. Reddy, N., Yang, Y., Potential of plant proteins for medical applications. *Trends in Biotechnol.*, 29, 490–498, 2011.
22. Surgical Dressing. http://emedicine.medscape.com/article/1127868-overview. Accessed 10th February 2015.
23. Sezer, A.D., Cevher, E., Biopolymers as wound healing materials: challenges and new strategies, biomaterials applications for nanomedicine, Prof. Rosario Pignatello, Ed., ISBN: 978-953-307-661-4, InTech, Available from: http://www.intechopen.com/books/biomaterials-applications-fornanomedicine/biopolymers-as-wound-healing-materials-challenges-and-new-strategies
24. Mackay, D., Miller, A.L., *Altern. Med. Rev.*, 8, 359, 2003.
25. Yoshie-Stark, Y., Wada, Y., Wasche, A., *Food Chem.*, 107, 32, 2008.
26. Thirugnanaselvam, M., Gobi, N., Arun Karthick, S., SPI/PEO blended electrospun martrix for wound healing. *Fiber. Polym.*, 14, 965–969, 2013.
27. Li, Y.-Y., *Fibers Text. East. Eur.*, 8, 12, 2004.
28. Ghorpade, V.M., Gennadios, A., Hanna, M.A., Weller, C.L., *Cereal Chem.*, 72, 559–563, 1995.
29. Shankar, A., Seyam, A.-F.M., Hudson, S.M., Electrospinning of soy protein fibers and their compatibility with synthetic polymers. *J. Textile Apparel Technol. Manag.*, 8, 1–14, 2013.
30. Har-el, Y., Gerstenhaber, J.A., Brodsky, R., Huneke, R.B., Lelkes, P.I., Electrospun soy protein scaffolds as wound dressings: enhanced reepithelialization in a porcine model of wound healing. Wound Med., 5, 9–15, 2014.
31. Har-el, Y., Brodsky, R., Gerstenhaber, J.A., Huneke, R.B., Lelkes, P.I., Electrospun protein scaffolds improve healing in a pig full thickness skin excision wound model. http://imgsvr.eventrebels.com/ERImg/00/91/88/2340042/28445-2-10311.pdf. Accessed 17th February 2015.

32. Peles, Z., Zilberman, M., Novel soy protein wound dressings with controlled antibiotic release: mechanical and physical properties. Acta Biomaterialia, 8, 209–217, 2012.
33. Peles, Z., Binderman, I., Berdicevsky, I., Zilberman, M., Soy protein films for wound-healing applications: antibiotic release, bacterial inhibition and cellular response. *J. Tissue Eng. Regen. Med.*, 7, 401–412, 2013.
34. Silva, S.S., Goodfellow, B.J., Benesch, J., Rocha, J., Mano, J.F., Reis, R.L., Morphology and miscibility of chitosan/soy protein blended membranes. *Carbohydr. Polym.*, 25–31, 2007.
35. Ramnath, V., Sekar, S., Sankar, S., Sastry, T.P., Mandal, A.B., *In vivo* evaluation of composite wound dressing material containing soya protein and sago starch. *Int. J. Pharm. Pharm. Sci.*, 4, 1–6, 2012.
36. Santos, T.C., Höring, B., Reise, K., Marques, A.P., Silva, S.S., Oliveira, J.M., Mano, J.F., Castro, A.G., Reis, R.L., van Griensven, M., *In vivo* performance of chitosan/soy-based membranes as wound-dressing devices for acute skin wounds. *Tissue Eng.: Part A*, 19, 860–869, 2013.
37. Eisenbud, D., Hunter, H., Kessler, L., Zulkowski, K., Hydrogel wound dressings: where do we stand in 2003? *Ostomy Wound Manage*, 49, 52–57, 2003.
38. Seaman, S., Dressing selection in chronic wound management. *J. Am. Podiatr. Med. Assoc.*, 92, 24–33, 2002.
39. Snyders, R., Shingel, K.I., Zabeida, O., Roberge, C., Faure, M.-P., Martinu, L., Klemberg-Sapieha, J.E., Mechanical and microstructural properties of hybrid poly(ethylene glycol)–soy protein hydrogels for wound dressing applications. *J. Biomed. Mater. Res., Part A*, 83A, 88–97, 2007.
40. Shingel, K.I., Stabile, L.D., Marty, J.-P., Faure, M.-P., Inflammatory inert poly(ethylene glycol)–protein wound dressing improves healing responses in partial and full-thickness wounds. *Int. Wound J.*, 3, 332–342, 2006.
41. Shevchenko, R.V., Santin, M., Pre-clinical evaluation of soybean-based wound dressings and dermal substitute formulations in pig healing and non-healing *in vivo* models. *Burns & Trauma*, 2, 187–195, 2014.
42. Ha, T.L.B., Quan, T.M., Vu, D.N., Si, D.M., Naturally derived biomaterials: preparation and application, Chapter 11, Regenerative medicine and tissue engineering edited by Jose A. Andrades, ISBN 978-953-51-1108-5, Published: 2013 http://www.intechopen.com/books/regenerative-medicine-and-tissue-engineering/naturally-derived-biomaterials-preparation-and-application.
43. Xu, H., Cai, S., Sellersc, A., Yang, Y., Intrinsically water-stable electrospun three dimensional ultrafine fibrous soy protein scaffolds for soft tissue engineering using adipose derived mesenchymal stem cells. *RSC Adv.*, 4, 15451–15457, 2014.
44. Wang, J., Ma, H., Jin, X., Hu, J., Liu, X., Ni, L., Ma, P.X., *Biomaterials*, 32, 7822–7830, 2011.
45. Silva, G.A., Vaz, C.M., Coutinho, O.P., Cunha, A.M., Reis, R.L., *In vitro* degradation and cytocompatibility evaluation of novel soy and sodium caseinate-based membrane biomaterials. *J. Mater. Sci. Mater. Med.*, 14, 1055–1066, 2003.

46. Chien, K.B., Aguado, B.A., Bryce, P.J., Shah, R.N., *In vivo* acute and humoral response to three-dimensional porous soy protein scaffolds. *ActaBiomater.*, 9, 8983–8990, 2013.
47. Chien, K.B., Shah, R.N., Novel soy protein scaffolds for tissue regeneration: Material characterization and interaction with human mesenchymal stem cells. *ActaBiomater.*, 8, 694–703, 2012.
48. Silva, S.S., Santos, M.I., Coutinho, O.P., Mano, J.F., Reis, R.L., Physical properties and biocompatibility of chitosan/soy blended membranes. J. Mater. Sci: Mater. Med., 16, 575–579, 2005.
49. Xu, X., Jiang, L., Zhou, Z., Wu, X., Wang, Y., Preparation and properties of electrospun soy protein isolate/polyethylene oxide nanofiber membranes. *ACS Appl. Mater. Interf.*, 4, 4331–4337, 2012.
50. Lin, L., Electrospun soy protein-based scaffolds for skin tissue engineering and wound healing, PhD thesis, Drexel University, 2011, https://idea.library.drexel.edu/islandora/object/idea%3A3437. Accessed 11th February 2015.
51. Santos, T.C., Marques, A.P., Silva, S.S., Oliveira, J.M., Mano, J.F., Castro, A.G., Reis, R.L., *In vitro* evaluation of the behaviour of human polymorphonuclear neutrophils in direct contact with chitosan-based membranes. *J. Biotechnol.*, 132, 218–226, 2007.
52. Ramji, K., Shah, R.N., Electrospun soy protein nanofiber scaffolds for tissue regeneration. *J. Biomater. Appl.*, 29, 411–422, 2014.
53. Chien, K.B., Makridakis, E., Shah, R.N., Three-dimensional printing of soy protein scaffolds for tissue regeneration. *Tissue Eng. Part C: Methods*, 19, 417–426, 2013.
54. Bhatia, S.K., Bio-based materials steps into the operating room. www.aiche.org/cep, 2012, 49–53. Accessed 18th February 2015.
55. Chien, K.B., Chung, E.J., Shah, R.N., Investigation of soy protein hydrogels for biomedical applications: Materials characterization, drug release, and biocompatibility. *J. Biomater. Appl.* 28, 1085–1096, 2014.
56. Santin, M., Ambrosio, L., Soybean-based biomaterials: preparation, properties and tissue regeneration potential, *Exp. Rev. Med. Devices*, 5, 349–358, 2008.
57. Burg, K.J.L., Porter, S., Kellam, J.F., Biomaterial developments for bone tissue engineering. Biomaterials, 21, 2347–2359, 2000.
58. Tiedeman, J.J., Garvin, K.L, Kile, T.A., Connolly, J.F., The role of a composite, demineralized bone matrix and bone marrow in the treatment of osseous defects. *Orthopaedics*, 18, 1153–1158, 1995.
59. Gepstein, R., Weiss, R.E., Hallel, T., Bridging large defects in bone by demineralized bone matrix in the form of a powder. A radiographic, histological, and radioisotope-uptake study in rats. *J. Bone Jt. Surg. Am.*, 69, 984–992, 1987.
60. Summers, B.N., Eisenstein, S.M., Donor site pain from the ilium. *J. Bone Jt. Surg. Br.*, 71-B, 677–680, 1989.
61. Coventry, M.B., Tapper, E.M., Pelvic instability. *J. Bone Jt. Surg. Am.*, 54-A, 83–101, 1972.
62. Younger, E.M., Chapman, M.W., Morbidity at bone graft donor site. *J. Orthop. Trauma*, 3, 192–195, 1989.

63. Bostrom, R.D., Mikos, A.G., Tissue engineering of bone. A. Atala, D Mooney, J. P. Vacanti, R Langer, Eds., *Synthetic biodegradable polymer scaffolds*, Birkhäuser, Boston (1997), pp. 215–234.
64. Vaz, C.M., Mano, J.F., Fossen, M., van Tuil, R.F., de Graaf, L.A., Reis, R.L., Cunha, A.M., Mechanical, dynamic-mechanical, and thermal properties of soy protein-based thermoplastics with potential biomedical applications. *J. Macromol. Sci. Part B Phys.*, 41, 33–46, 2002.
65. Santin, M., Morris, C., Standen, G., Nicolais, L., Ambrosio, L., A new class of bioactive and biodegradable soybean-based bone fillers. *Biomacromolecule*, 8, 2706–2711, 2007.
66. Meikle, S.T., Standen, G., Salvage, J., Santis, R.D., Nicolais, L., Ambrosio, L., Santin, M., Synthesis and characterization of soybean-based hydrogels with an intrinsic activity on cell differentiation. *Tissue Eng. Part A*, 18, 1932–1939, 2012.
67. Perut, F., Montufar, E.B., Ciapetti, G., Santin, M., Salvage, J., Traykova, T., Planell, J.A., Ginebra, M.P., Baldini, N., Novel soybean/gelatine-based bioactive and injectable hydroxyapatite foam: Material properties and cell response. Acta Biomaterialia, 7, 1780–1787, 2011.
68. Merolli, A., Nicolais, L., Ambrosio, L., Santin, M., A degradable soybean-based biomaterial used effectively as a bone filler *in vivo* in a rabbit. *Biomed. Mater.*, 5, 15008, 2010.
69. Giavaresi, G., Fini, M., Salvage, J., Aldini, N.N., Giardino, R., Ambrosio, L., Nicolais, L., Santin, M., Bone regeneration potential of a soybeanbased filler: Experimental study in a rabbit cancellous bone defects. *J. Mater. Sci. Mater. Med.*, 21, 615–626, 2009.
70. Silva, R.M., Elvira, C., Mano, J.F., San Román, J., Reis, R.L., Influence of beta-radiation sterilisation in properties of new chitosan/soybean protein isolate membranes for guided bone regeneration. *J. Mater. Sci. Mater. Med.*, 15, 523–528, 2004.
71. Silva, R., Bulut, B., Roether, J.A., Kaschta, J., Schubert, D.W., Boccaccini, A.R., Sonochemical processing and characterization of composite materials based on soy protein and alginate containing micron-sized bioactive glass particles. J. Mol. Struct., 1073, 87–96, 2014.
72. Chen, L., Subirade, M., Food-protein-derived materials and their use as carriers and delivery systems for active food components. In: *Delivery and controlled release of bioactives in foods and nutraceuticals*, N. Garti, Ed., Woodhead Publishing Ltd, UK (2008), pp. 251–278
73. Panyam, K.D., Peptides from milk proteins and their properties. *Crit. Rev. Food Sci. Nutr.*, 43, 607–633, 2003.
74. Korhonen, H., Pihlanto, A., Bioactive peptides: new challenges and opportunities for dairy industry. *Aust. J. Dairy Technol.*, 58, 129–134, 2003.
75. Chen, L., Remondetto, G., Rouabhia, M., Subirade, M., Kinetics of the breakdown of cross-linked soy protein films for drug delivery. Biomaterials, 29, 3750–3756, 2008.

76. Vaz, C.M., van Doeveren, P.F.N.M., Reis, R.L., Cunha, A.M., Soy matrix drug delivery systems obtained by melt-processingtechniques. *Biomacromolecules,* 4, 1520–1529, 2003.
77. Vaz, C.M., van Doeveren, P.F.N.M., Reis, R.L., Cunha, A.M., Development and design of double-layer co-injection moulded soy protein based drug delivery devices. *Polymer,* 44, 5983–5992, 2003.
78. Siva Mohau Reddy, G., Jayaramudu, J., Sadiku, E.R., Shina Ray, S., Varaprasad, K., Aderibigbe, B.A., Application of cross-linked soy protein isolate with resorcinol films for drug release studies of naturally occurring bioactive agent with antiproliferative activity, *J. Drug Del. Sci. Tech.*, 24, 86–91, 2014.
79. Nayak, P., Sahoo, S.K., Behera, A., Nanda, P.K., Nayak, P.L., Guru, B.C., Synthesis and characterization of soy protein isolate/mmt nanocomposite film for the control release of the drug ofloxacin. *World J. Nano Sci. Eng.*, 1, 27–36, 2011.
80. Sasmal, A., Nayak, P., Nanda, R., Nayak, P.L., Sasmal, S., Chang, Y.-W., Kang, S.C., Yoon, J.-Y., Soy protein isolate – furfural cross-linked nanocomposites for controlled release of cefadroxil. *Int. J. Plast. Technol.*, 13, 8–21, 2009.
81. Pradhan, N.K., Das, M., Palve, Y.P., Nayak, P.L., Synthesis and characterization of soya protein isolate/cloisite 30B (MMT) nanocomposite for controlled release of anticancer drug curcumin. *Int. J. Res. Pharm. Biomed. Sci.*, 3, 1513–1522, 2012.
82. Vega-Lugo, A.-C., Lim, L.-T., Controlled release of allylisothiocyanate using soy protein and poly(lactic acid) electrospun fibers, *Food Res. Int.*, 42, 933–940, 2009.
83. He, W., Tan, Y., Tian, Z., Chen, L., Hu, F., Wu, W., Food protein-stabilized nanoemulsions as potential delivery systems for poorly water-soluble drugs: preparation, *in vitro* characterization and pharmacokinetics in rats. *Int. J. Nanomed.*, 6, 521–533, 2011.
84. Liu, Y., Cui Y., Liao, M., pH and temperature-responsive IPN hydrogels based on soy protein and poly(*N*-isopropylacrylamide-*co*-sodium acrylate). *J. Appl. Polym. Sci.*, 131, 39781, 2015.
85. Yin, B., Deng, W., Xu, K., Huang, L., Yao, P., Stable nano-sized emulsions produced from soy protein and soy polysaccharide complexes. *J. Colloid Interface Sci.*, 380, 51–59, 2012.
86. Maltais, A., Remondetto, G.E., Subirade, M., Tabletted soy protein cold-set hydrogels as carriers of nutraceutical substances. Food hydrocolloid., 24, 518–524, 2010.
87. Liu, Y., Wei, S., Liao, M., Drug release mechanism of fast temperature-responsive soy protein/PNIPAAm IPN hydrogels. *Adv. Mater. Res.*, 781–784, 803–807, 2013.
88. Liu, Y., Cui, Y., Preparation and properties of temperature-sensitive soy protein/poly(N-isopropylacrylamide) interpenetrating polymer network hydrogels. *Polym. Int.*, 60, 1117–1122, 2011.

6

Electrospinning of Soy Protein Nanofibers: Synthesis and Applications

Carlos L. Salas

Department of Forest Biomaterials, North Carolina State University, Raleigh, USA
Currently at *Georgia-Pacific Corporation. Innovation Institute. Neenah, Wisconsin. USA*

Abstract
Natural materials have been in recent scrutiny from the scientific community for their potential use in a variety of industrial applications. Among these natural polymers, proteins from different sources (wheat, silk, casein, soybean, and others) have been used in preparation of composite materials. Soybean proteins are present in products commercially available such as soy flours, soy concentrates, and soy isolates, each of these products with different protein content and functional properties. Soy proteins have found industrial use in wood adhesives, composite and building materials, textile fibers, inks and plastics. This chapter presents a summary of recent research on the use of electrospinning technique to produce nanofibers from soybean proteins. Electrospinning has been used to produce nanofibers from different synthetic and natural polymers, these nanofibers have high surface area and offer numerous advantages to use them in different applications. The chapter starts with a brief discussion on the physical properties of soy proteins that affect their electrospinning into nanofibers, followed by a summary of the relevant literature about electrospinning of soy proteins and finally the chapter ends with a discussion on the literature about prospect applications of electrospun nanofibers from soybean proteins.

Keywords: Soy proteins, electrospinning, electrospun nanofibers, natural polymers

Corresponding author: clsalasa@ncsu.edu

Visakh P. M and Olga Nazarenko (eds.) Soy Protein-Based Blends, Composites and Nanocomposites, (135–154) 2017 © Scrivener Publishing LLC

6.1 Introduction

Soybeans are an important crop worldwide. Once the oil has been extracted from the beans, the protein meal is further processed into products having different properties and different protein content, namely soy flour, soy concentrate, and soy isolates, with protein contents estimated to be in the range of 56%, 65%, and 90%, respectively. These protein-rich products have found uses in many non-food industrial applications, including the manufacture of plastics, adhesives, paper coatings, paint coatings, and composites. The advantages of using soy proteins in such systems are not only due to their physicochemical properties but also because of their renewability and sustainable production.

Research efforts in the last decade have pointed toward the use of bio-based materials in different applications as an aim to reduce the consumption of synthetic polymers, which can be harmful to the environment. As such, the use of natural, renewable resources in nanotechnology applications offers unique advantage. Soybean proteins alone or in combination with other natural and synthetic polymers have been used to produce nanofibers by the electrospinning technique. In the solution electrospinning process, a polymer is first dissolved in a given solvent and the solution is pumped through a nozzle that together with a metallic fiber collector become the electrodes between which an electric field is applied. The electrical field stretches the polymer solution at the tip of the capillary, forming a conical shape, the Taylor cone. The fluid is then expelled from the nozzle as a thin jet and the solvent evaporates, forming the fibers. The nanofibers produced by electrospinning have a high surface area, which makes them suitable for different applications, including textiles, tissue engineering, cosmetics, wound dressing, filtration, catalysis, and etc.

This chapter provides an overview of the electrospinning of soybean proteins to produce nanofibers, the variables that have been reported to affect the electrospinning of soy proteins and some potential applications.

6.2 Properties of Soybean Proteins That Affect Electrospinning

Soybeans contain different proteins, the storage proteins, which serve as nitrogen storage for the plant, and proteins with biological activity, namely lipoxygenases and tripsin inhibitor [1]. The storage protein in soybeans accounts for a large fraction of the raw bean weight (between 65% and 80%) [2]. Storage proteins are globulins, that is, their solubility in water is enhanced by the presence of electrolytes [2]. They have been classified according to the sedimentation constant as 7S and 11S or βconglycinin and glycinin, respectively. The 7S fraction (β conglycinin) has a molecular weight of 180–200 kDa and is composed of three subunits, α (57–76 kDa), α' (57–83 kDa), and β (42–53 kDa). The 11S fraction (glycinin) has a molecular weight of 320–350 kDa and its quaternary structure is formed by 12 subunits forming a dimer. The isoelectric points of soy globulin 7S and 11S are 4.96 and 4.64, respectively [2–4].

The use of soy proteins is closely related to their physicochemical properties including solubility, hydration properties, gelation and interfacial activities, which, in turn, are governed by the structure and charge balance of the macromolecule.

It has been reported that heating soybean protein solutions also results in the formation of soluble aggregates; however, this thermal denaturation is sensitive to the ionic strength and the addition of disulfide reducing agents. The addition of salt increases denaturation temperature, for instance, at 0.001 M ionic strength and heating at 60 °C, the amount of 11S soluble decreased (quantified in supernatant after centrifugation), whereas at ionic strength of 0.1 M, the protein remained soluble [3].

Both of the storage proteins can form disulfide linkages; however, glycinin units have a bigger amount of half cysteine per mole of protein (48 mol/mol) [3] compared to 7S globulins (4 mol/mol) [4]. This affects their physical properties and the solubility to a greater extent. 11S protein forms reversible association polymers at pH 7.5 when ionic strength decreases from 0.5 to 0.1. It has been reported that at higher pHs, the 11S protein contains more SH groups compared to pH 7.6, a value that increases with the unfolding of the protein, which indicates that most SS groups are buried inside the protein molecules and the alkaline treatment cleaves it from SS to SH. Reducing agents such as 2-mercaptoethanol, cysteine, NaCN, and dithiothreitol (DTT) have been used to break disulfide bonds in soy protein. Glycinin contains 2 free mol of sulfhydryl group/mol protein in its native state and 2–3 mol of sulfhydryl/mol after heating [3].

Regarding the secondary structure, both proteins show a high content of random coil structure and a low content of ordered structure. For instance, 7S protein contains a 5% α-helix, 35% β structure, and 60% random coil, 11S protein shows a secondary structure with 5.2% α-helix, 34.8% β structure, and 60% random coil [5]. This native structure has been shown to change with alkali treatment (which decreases ordered structures) or alkali-alcohol treatment, which increases ordered structures (α-helix) [3].

Although the pure soy protein fractions are of interest to evaluate their contribution to the whole behavior of the proteins, commercially available soy protein products are not pure fractions but rather contain both storage proteins. Therefore, in industrial applications, the focus is in the utilization of commercial soy protein products, similarly in the case of electrospinning.

As a first step, the solution electrospinning process requires solubilization of the polymer of interest, which requires appropriate selection of the solvent. As soy proteins are globulins, their solubility in water is enhanced by the content of electrolytes. In addition, the pH will affect the solubility. The isoelectric pH of soy proteins have been reported as 4.5 [2, 3]. The solubility will be low to zero at pH values near the isoelectric pH and increased at higher pHs. Some alternatives to enhance the solubility in aqueous solvents involve the use of chaotropic agents (urea, guanidine hydrochloride), which disrupt the hydrogen bonding and dissolve the protein; however, the use of such agents can leave a significant amount of salt after solvent evaporation, an undesirable effect for most applications.

Other polar solvents have been also explored to solubilize soy proteins for electrospinning purposes. For instance, Lin [6] studied different solvents for dissolving SPI to obtain solutions suitable for electrospinning. The solvents included water, acetic acid, ethanol, hydrochloric acid, acetone, sodium hydroxide, ammonium hydroxide, and some polar but less water soluble solvents, namely dimethylformamide (DMF), tetrahydrofuran (THF), and 1,1,1,3,3,3 Hexafluoro-2-propanol (HFIP). In the study, only HFIP proved to be a stable solvent for soy protein to produce uniform electrospun fibers, whereas the other solvents only produced electrospray [6].

The search for a suitable organic solvent for soy proteins involves the understanding of the role of protein-solvent interactions. Although protein structure resembles that of synthetic polymers in some respects, the protein structure is more complicated due to the different functionality of the constituent aminoacids, some hydrophilic, and some hydrophobic. This difference in functionality is what confers proteins their amphiphilic nature and also plays an important role on the interactions with the solvent.

One way to approach the problem of finding a good organic solvent for soy proteins could be using the solubility theories that are applied to synthetic polymers, for instance, the use of the solubility parameter.

The Hansen solubility parameter [7] includes a modification of the cohesive energy (E) to account individually for all the molecular interactions in a mole of material, namely dispersion forces (E_D), polar interactions (dipole-dipole, E_P), and specific interactions such as hydrogen bonding (E_H).

$$E = E_D + E_P + E_H \tag{6.1}$$

which after dividing by the molar volume (V) are:

$$\frac{E}{V} = \frac{E_D}{V} + \frac{E_P}{V} + \frac{E_H}{V} \tag{6.2}$$

The Hansen solubility parameter is defined as:

$$\delta^2 = \delta_d^2 + \delta_p^2 + \delta_h^2 \tag{6.3}$$

δ is the Hansen solubility parameter, δ_d represents the dispersive component, δ_p the polar, and δ_h the hydrogen bonding. The cohesive energy can be expressed as

$$E = \Delta H - RT \tag{6.4}$$

where ΔH refers to the latent heat of vaporization, T is the absolute temperature, and R is the universal gas constant [7].

In terms of Hansen solubility parameter, a good solvent for a certain polymer has a solubility parameter close to that of the polymer. One way to identify a good solvent for a polymer is by calculating the Hansen solubility parameter distance, which is given for two substances by:

$$Ra^2 = 4\left(\delta_{d_2} - \delta_{d_1}\right)^2 + \left(\delta_{p_2} - \delta_{p_1}\right)^2 + \left(\delta_{h_2} - \delta_{h_1}\right)^2 \tag{6.5}$$

The smaller the value of Ra, the more solvent and solute like each other. In addition, the relative energy difference, defined as RED = Ra/R_o, where R_o is the radius of a sphere having the three components of Hansen solubility parameter as coordinates. For instance, good solvents will have RED values less than one [7].

Recently, Aghanoury and Sun [8] studied the efficacy of 100 different solvents and solvent mixtures to solubilize soy proteins using Hansen solubility parameter. They used the storage proteins in soybean (glycinin and β-conglycinin). Their approach consisted of identifying a solvent system that matches the Hansen solubility parameter of urea solutions, in which soy protein dissolves very well. They used the solubility parameter distance Ra, as an indicator of the suitability of a solvent to dissolve soy protein. To this end, 2% soy protein solutions were prepared in each solvent, and the solubility was then measured using a protein assay method (i.e., determination of nitrogen content). High solubilization (>80% glycinin or β-conglycinin) was observed for solvent mixtures containing water-triethanolamine and 1,4-dioxane or 2-pyrrolidone [8]. This report illustrates the applicability of a practical concept on the determination of a suitable solvent for soy proteins. However, pure soy protein globulins are rarely used in practical applications due to the cost involved in purification. Furthermore, the generalization is not straightforward, and the selection of a "suitable" solvent must fulfill the requirements of dissolving the solute within a reasonable time, being good solvent and being environmentally friendly as well. For instance, if we calculate the *Ra* value for 1,1,1,3,3,3-hexafluoroisopropanol, for which components of the Hansen solubility parameter are δ_d = 17.2, δ_p = 4.5, and δ_h = 14.7 [7] (considering the Hansen solubility parameters of urea 6M as those of the protein), it gives *Ra* of 26.3.This high value would indicate that the HFIP solvent is not a good solvent for soy proteins; however, it has been used to solubilize soy proteins for electrospinning. Furthermore, the studies that report use of HFIP [6,9–11] indicate dissolution times of as low as 48 h and up to 2 weeks. This could be a limitation for cost-effective applications and scale-up to produce electrospun fibers. On the other hand, the use of pure protein fractions could be also of little practical interest. That is why most reports have focused on the use of commercially available soy protein. Typically isolate has been used because of its high protein content; however, its cost is still high compared to soy flour that has been used very recently [12, 13] on electrospinning applications.

It is also worth mentioning that the native state of storage proteins in soybeans and commercial products includes the two molecules linked by disulfide bonds, which also affects solubility and physical properties. The use of reducing agents can break apart these linkages and allows for molecular unfolding. An attempt to unfold the protein structure of soybean protein solutions for electrospinning using cysteine was recently reported [14]. Alkaline solutions (pH 10, 11) have also proved useful in the unfolding of the globular structure of soy proteins. Moreover, the use of small amounts of surfactant improved the spinnability of soy protein solutions. The nonionic surfactant Triton X-100 [12,13,15–18] and anionic sodium dodecyl sulfate [14,19] have been used, from concentrations as small as 0.5 wt% to as high as 17.5 wt%. Surfactants can serve a twofold purpose: as coadjutants in the unfolding/denaturation of soy protein molecules and to reduce surface tension of the electrospinning solutions.

Additional complications in term of solubility and preparation of solutions for electrospinning arise when soy protein are blended with other polymers. This is because the solvent of choice has to be a good solvent for both the protein and the coadjutant polymer. In addition, the molecular weight of the co-spinning polymer will affect the viscosity and spinnability of the blend. Different coadjutant polymers have been used for electrospinning of soybean proteins from aqueous solutions. This implies that the coadjutant polymer should be water soluble. To this end, polyethylene oxide and polyvinyl alcohol have been used. Other biopolymers include corn zein [20], wheat protein (gluten) [13], and lignin [21]. One disadvantage of the use of water soluble polymers is the hydrophilicity of the obtained electrospun fibers. This can be improved by blending soy protein with more hydrophobic polymers; however, the solvent of choice is again the main issue. For instance, in the electrospinning of soy protein-PCL blends, formic acid, and acetic acid where used [22]. A list of the solvents and systems used for electrospinning of soy protein is given in Table 6.1.

6.2.1 Electrospinning of Soybean Proteins

The first reports on the electrospinning of soy protein date back to the works of Phiriyawirut *et al.* [20] and Vega-Lugo and Lim [15] in 2008. For instance, Phiriyawirut *et al.* [20] tried to produce electrospun nanofibers from soy protein isolate by dissolving it to 5 wt%, 6 wt%, and 7 wt% in either formic or acetic acid. The results indicated high viscosity of the protein solution formic acid (134.1 cP) vs acetic acid (29.8 cP). However, at the studied conditions, the soy isolate alone did not produce fibers in any of the studied solvents, only beads. It was shown, however, that the diameter of beads formed decreased with increasing SPI concentration. It was necessary to blend soy protein solutions in formic or acetic acid with zein (25% zein dissolved in ethanol). The blend containing 95% zein and 5% soy protein (on dry basis) was able to produce uniform fibers, with better results when SPI was dissolved in acetic acid (5% in acetic acid) compared to formic acid. The effect of applied voltage (20–28 kV) and distance to collector were also studied, with fibers of smaller diameter obtained at higher voltages. Altering the distance to collector did not change the morphology of fibers. Shorter distances favored junction (collapse of the fibers). Although the fibers were uniform, the blends with zein reduced drastically the content of soy protein on the final fibers, that is, the fibers were zein fibers containing soy protein [20].

The study carried out by Vega-Lugo and Lim [15] found again that the soy protein isolate alone did not produce any fibers, even after thermal treatment (60 °C, 2 h) and dissolution in 1% NaOH solution. Afterwards, SPI solutions with content up to 15 wt% SPI and different amounts of PEO (900 kDa, 0.2 and 0.8 wt%) and Triton X surfactant (1 wt%) were prepared. The results showed that the PEO produced both an increase in viscosity and a reduction in thermal conductivity of the precursor solutions, with uniform fibers obtained at PEO concentration of 0.8 wt%. It was also found that the solutions of SPI with viscosity less than 10 cP did not produce fibers. Uniform fibers were obtained from solutions with viscosity higher

Table 6.1 Summary of solvents and main conditions used in different soy protein electrospinning studies.

System	Solvent	Coadjutant polymer (Mw)	Other coadjutant molecules	Thermal treatment	Voltage	Distance to collector (cm)	Size of nanofibers	Reference
SPI/zein	Acetid acid, Formic acid	Zein (25 to 29kDa)	None	None	20–28 kV	10–16	-	[20]
SPI/PEO	1 wt% NaOH-Water	PEO (900 kDa)	Triton X-100, 1 wt%	60 °C, 2 h	20–30 kV	26	200–260 nm	[15]
SPI	1 wt% NaOH-Water	PEO (900 kDa)	Triton X-100 1 wt%, cyclodextrin	60 °C, 2 h	20–30 kV	26	188–266 nm	[16]
SPI/PVA	Water, 3 different pH	PVA, (78 kDa)	Triton X-100	60, 80, 95 °C. 30, 60 min, pH 7,9,12	16 kV	15	0.6–4.5 µm	[18]
SPI/PEO	HFIP, NaOH	PEO (1000 kDa)	None	None	15 kV	12	250–500 nm	[6]
SPI/PVA	Aqueous solvent (3 different pH and thermal treatment)	PVA (78 kDa)	Triton X-100, 0.5 wt%	80 °C, 30 min	15 kV	15	0.4–0.8 mm	[17]
SPI/PVA	Alkaline solution, pH 11	PVA (1750 Da)	Sodium dodecyl sulfate (0.6 wt%)	60 °C, 2 h	12–14 kV	10–15	200–300 nm	[19]
SPI/PEO	1,1,1,3,3,3-hexafluoro-2-propanol (HFIP)	PEO (200kDa)	None	None	25 kV	25	200–300 nm	[9]
SPI/PEO	HFIP	PEO (1000 kDa)	None	None	12 kV (SPI/PEO) 20-25 kV (SPI/zein)	15	700–1250 nm	[10]
SPI/PEO	SPI in 1 wt% NaOH and PEO in ethanol	PEO (300 kDa)	None	None	15 kV	15	151–415 nm	[23]
SPI/PEO	1 wt% NaOH	PEO (900 kDa)	None	60 °C, 2 h	20–30kV	20	-	[25]

SPI/PEO	HFIP	PEO (1000 kDa)	None	None	12 kV	15	–	[11]
Soy flour/ PEO	Alkaline (NaOH) solutions, pH 11.	PEO (600 kDa)	Triton X (1 wt%)	None	30 kV	15	–	[12]
SPI/PEO	1 % NaOH	PEO (100 kDa, 1000 kDa, 8000 kDa)	None	60 °C, 1 h	15–27 kV	18	50–270 nm	[24]
SPI/PEO	0.1N NaOH/acetonitrile	PEO (400 kDa)	Lignin	None	15 kV	22	Varies, 110–500 nm	[21]
SPI	Water/cysteine (10 wt%)/ SDS	None	Cysteine/SDS	None	45 kV	25	–	[14]
Soy flour	Alkaline solution, pH 11	PVA (130 kDa)-gluten	Triton X-100, 0.5 wt%	None	25 kV	15	–	[13]
SPI/PVA and SPI/PCL	Formic acid, acetic acid. Ethanol/water	PVA (9 kDa, 130 kDa), PCL(80 kDa)	Water soluble adhesives	None	15-18 kV (PVA), 14-16 kV (PCL)	10–15	700–800 (SP/ PVA), 1000 (PCL)	[22]

Abbreviations: SPI – soy protein isolate; PEO – Poly(ethylene oxide); PVA – poly(vinyl alcohol); PCL – polycaprolactone; HFIP – hexafluoroisopropanol; SDS – sodium dodecyl sulfate.

than 25 cP. The best morphology was achieved with a ratio 95/5 SPI/PEO. All commercial soy isolate samples studied were able to produce fibers at concentrations of 15% SPI, 0.6% PEO, 0.5% surfactant concentration. In addition, changing the polarity of the spinneret produced fibers with lower diameter [15].

Cho *et al.* [18] studied the effect of processing conditions on fiber formation and properties. Their approach included the preparation of soy protein isolate (8 wt%, 80 °C) and polyvinyl alcohol (15 wt%, Mw = 78 kDa) solutions separately prior to mixing which improved spinnability. They screened three different SPI/PVA ratios prior to selecting a total polymer concentration of up to 13% (50:50 SPI:PVA). Further evaluation of the 50:50 mixture included assessment of three different pH values (7, 9, and 12) and three different temperatures to denature the SPI (60 °C, 80 °C, and 95 °C) at different heating times. The addition of Triton X surfactant up to 0.6 wt% helped to reduce surface tension and improve spinnability. The results indicate a decrease in viscosity of solutions with increase in pH, an effect attributed to the denaturation and unfolding of the protein molecules, which in turn reduced the molecules entanglement. Same decrease in viscosity was observed with temperature increase at constant pH. The diameters of the fibers increased with the polymer content at the same pH. The addition of PVA increased the viscoelasticity of SPI solution, in addition for a constant ratio SPI:PVA the diameter of electrospun fibers increased with final total polymer concentration (see Figure 6.1).

A follow-up study of the same system [17] included the effect of changing the pH (9 and 12) on mechanical properties and biodegradability of soy protein isolate-polyvinyl alcohol electrospun fibers. SPI and PVA (78 kDa) in ratios of 15:85, 25:75, 35:65 with 0.5 % Triton X were studied. The

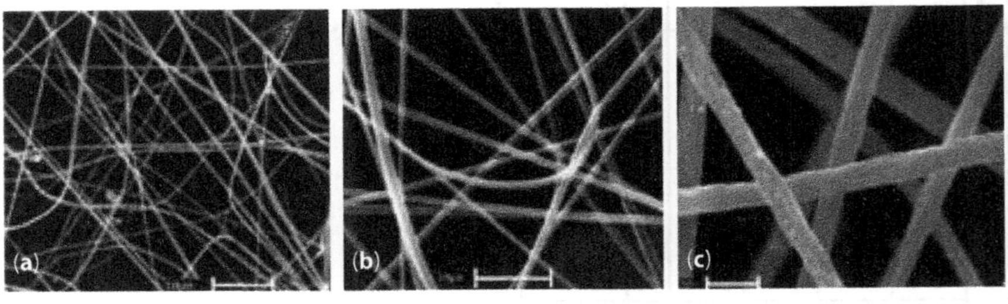

Figure 6.1 Electrospun soy protein/poly(ethylene oxide) from solutions with total polymer concentration of (a) 9 wt%, (b) 11 wt%, and (c) 13 wt%. The scale bar represents 10 μm. Figure reproduced from Cho *et al.* [18] with permission. Copyright © 2010 John Wiley & Sons, Inc.

electrospinning setup included the use of two collectors (flat and a cylindrical). The results indicate that at the same polymer content the higher pH promoted formation of thinner fibers which affected mechanical strength. Mechanical properties of SPI/PVA electrospun nanofibers mixtures were evaluated. Fibers with diameters ranging from 0.4 to 0.8 µm were obtained. Higher diameter was observed at lower pHs. The nanofiber mats prepared from solutions at pH 9 exhibited a higher load and elongation at break than those prepared at pH 12. This effect was ascribed to the lower denaturation of the proteins at pH 9 compared to pH 12. Similarly, the increase in concentration of SPI on the fiber mats produced a reduction in mechanical strength of the fibers. Further analysis included the composting of the fiber mats, the results indicate that the weight exhibited an exponential weight decay over time (up to 30 days), with a higher weight loss increasing with the content of the soy protein.

Hybrid SPI/PVA/SiO_2 nanofibers were prepared by addition of silicon dioxide (SiO_2) nanoparticles (up to 15 wt%) to SPI/PVA blends (8:2) at pH 10 (19). Different electric fields were used (0.75 kV/cm, 0.90 kV/cm, and 1.05 kV/cm), which produced fibers with diameters between 250 nm and 280 nm. Larger electric field strength, however, produced uneven fibers of about 110 nm. The nanofiber diameter decreased with the amount of SiO_2, in addition the higher concentration of SiO_2 particles promoted the formation of beads. More uniform fiber diameters were obtained when 0.6 wt% sodium dodecyl sulfate was added to the precursor solution, an effect that was ascribed to both the reduction on surface tension of dope and also to the enhancement in dispersion of SiO_2 particles. FTIR analysis of the hybrid fibers showed enhancement of hydrogen bond interactions. The fibers exhibit improved thermal stability with the addition of SiO_2 particles. Furthermore, the presence of SiO_2 particles reduced the crystallinity of PVA in the blend as shown by XRD analyses [19].

Xu *et al.* [9] produced SPI-PEO electrospun fibers with different SPI content (94%, 90%, 85%, and 67%). The solutions were prepared in 1,1,1,3,3,3-hexafluoro-2-propanol. Overall, the results indicated good miscibility between PEO and SPI as indicated by the uniform fiber formation obtained as the SPI content reduced on the samples. The fibers prepared with higher content (94%) of SPI did not have uniform fiber diameter and showed considerable branching, that is, small fibril branches attached to the bigger diameter fibers. The branched fibers accounted for 97% of the fiber population and had diameters of 195 nm. However, the branching disappeared as the SPI concentration in the fibers was decreased. X-ray diffraction results showed that the presence of SPI decreased the crystallization of PEO. The thermal stability (decomposition temperature) of the nanofibers was slightly lower than that of pure SPI, with faster decomposition rate. All the nanofiber samples exhibited low contact angles (high wettability), which could limit their practical application [9].

The synergy between soy proteins with other natural polymers to produce electrospun nanofibers has remained essentially unexplored, with only blends of zein-soy protein isolate [20], soy flour-gluten [13], and soy protein isolate-lignin [21] being reported. For instance, polysaccharides such as cellulose or starch and other wood derived polymers when combined with proteins could produce materials with interesting properties. Salas *et al.* [21] studied the synergy between soy proteins and lignin to produce electrospun nanofibers using 10% PEO (on a dry basis) as coadjutant polymer. Blends with different ratios of either soy glycinin or soy protein isolate and kraft lignin at a total polymer concentration of 8 wt% were prepared in alkaline solution (0.1 M NaOH, 10 % vol acetonitrile). The spinnability of soy glycinin and soy protein isolate was very similar. Overall, the results indicate good synergy between soy proteins and lignin with PEO as coadjutant. The addition of lignin to the protein solutions produced an increase in conductivity and a decrease in surface tension, which translated into better spinnability and more uniform fibers, with diameters ranging from 110 nm to 440 nm. The results also indicated higher viscosity for the solutions containing soy isolate compared to those with pure glycinin which in turn produced fibers with higher diameter. For both pure glycinin and soy isolate, the addition of lignin increased the fiber diameter (see Figure 6.2). Thermal analysis results indicated good compatibility of the fibers components and PEO [21]. One interesting point to highlight in terms of materials applications is the hydrophilicity of the final fibers, which suggests the need of cross-linking to improve their water resistance.

Because of its high protein content, soy protein isolate has been the product of preference in most electrospinning of soy protein reports; however, the use of soy flour was reported recently [12, 13]. Deffated soy flour that was subjected to an acid wash to increase its protein content was blended with gluten and PVA to produce fibers by electrospinning [13]. The blends were produced by mixing aqueous PVA solution (14 wt%) with soy protein/gluten mixtures that have been heated to 60 °C for 30 min, in addition, Triton X surfactant was used to improve spinnability. A "green" cross-linking agent prepared from oxidation of sucrose (OS) with hydrogen peroxide was compared against glyoxalfor the purpose of preparing water-resistant fibers. The cross-linking agent was added

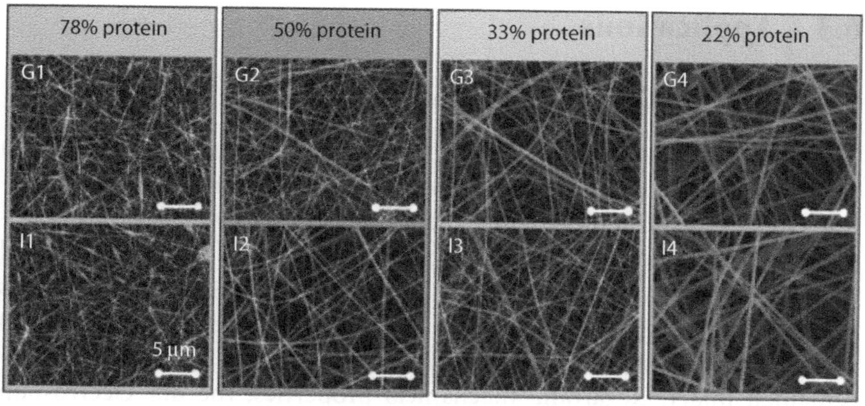

Figure 6.2 FE-SEM images of soy protein/lignin with 10 wt% (based on dry fibers) of poly(ethylene oxide) as coadjutant polymer. Top row: glycinin (G)–lignin (L) fibers with different protein/lignin content (G1, G2, G3, G4). Bottom row: soy isolate (I)-lignin (L) fibers (I1, I2, I3, I4). Fiber diameter increase with lignin content (from left to right).The scale bar shown is 5 μm [21].Reprinted from *Reactive and Functional Polymers*, Vol. 85, Carlos Salas, Mariko Ago, Lucian A. Lucia, Orlando J. Rojas, Synthesis of soy protein–lignin nanofibers by solution electrospinning, Pages 221–227., Copyright © 2014 with permission from Elsevier.

to the PVA solution prior to mixing, and the cross-linking reaction was completed by treatment of the nanofibers at 100 °C for 30 min. Uniform fibers were obtained from blends containing 30 wt% gluten, 25% purified soy flour, and 45% PVA on a dry basis. The results indicate that OS cross-linker performed similar to glyoxal. A high amount of nanofibers (>78%) remained insoluble up to 30 days at concentrations of OS of 10% and 15%. The SEM imaging of the fibers indicates that the nanofibers swell considerably when in contact with water. However, the fiber mats retained their structure [13]. Similar swelling behavior of electrospun nanofibers prepared from solutions of HFIP (without cross-linker) was reported by Lin *et al.* [6].

Soy protein/PVA (9 kDa and 130 kDa) and soy protein/PCL (80 kDa) fibers were electrospun on top of a rayon support membrane. The produced nanofibers were prepared with enhanced adhesive properties by either adding an adhesive solution as a coating layer on the electrospun fibers or adding the adhesive as part of the formulation of the doping solutions, in which case a surfactant was added to the solution. These adhesives help the electrospun fibers to stay on the rayon support.

Other studies of electrospinning of soy protein will be discussed next in terms of possible applications.

6.3 Applications

6.3.1 Wound Dressing Applications

Soy protein isolate solutions (up to 8wt% SPI) with 0.05% PEO in 1,1,1,3,3,3-hexafluoro-2-propanol were used to prepare nanofibers scaffolds by electrospinning [10] and then these scaffolds were used to evaluate cell growth (human dermal fibroblasts). The fibers were uniform and their size varied from 700 nm (5% SPI) up to 1250 nm (8% SPI). The results also indicate a high swelling of soy fibers (twofold increase in diameter) after immersing in phosphate buffer for 2 h. However, the morphology remained intact. Regarding mechanical properties, the dry electrospun scaffolds exhibited an ultimate tensile strength of ~1 MPa, which did not change significantly with concentration of SPI. However, when hydrated, the ultimate tensile strength dropped to 0.1 MPa. These nanofibers served to support growth human dermal fibroblasts cells after 8 days, and the results indicated that soy protein scaffolds provided a good substrate for adhesion and proliferation of cells (see Figure 6.3), which makes them good candidates for wound dressing applications [10].

Three-dimensional (3D) and regular (2D) soy protein scaffolds were prepared from soy protein solutions (up to 32.5 wt%) denatured with sodium dodecyl sulfate (SDS, 17.5 wt%) and cysteine as disulfide reducing agent [14]. The addition of sodium dodecyl sulfate and cysteine helps to

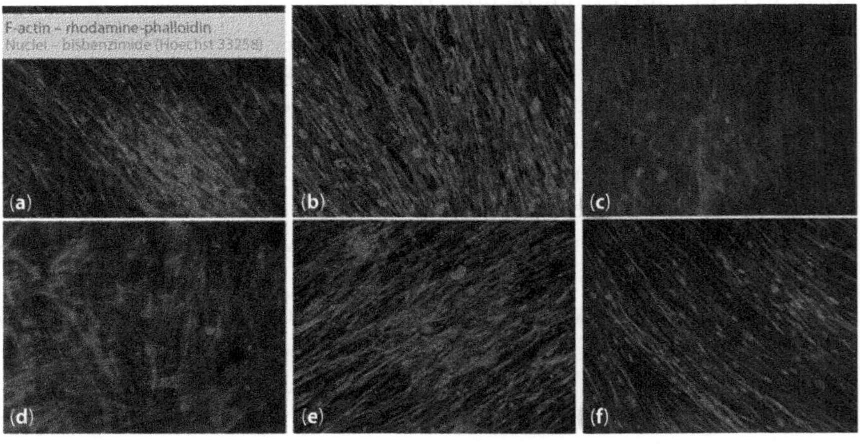

Figure 6.3 Growing of human dermal fibroblasts after eight days of culture on electrospunscaffolds prepared form (a) 5% SPI, 0.05% PEO, (b) 8% SPI, 0.05% PEO (c) 40% zein, (d) gelatin, (e) poly(lactic-co-glycolic acid), PLGA, and (f) glass. (Nuclei, blue; F-actin cytoskeleton, red. Magnification: 200x). Figure reproduced from Lin et al. [10] with permission. Copyright © 2013 John Wiley & Sons, Inc. [10].

solubilize the proteins. The solutions were thermally treated at 90 °C before electrospinning. The SDS page results indicate that the molecular weight of protein subunits remained the same after SDS and cysteine treatment, whereas those samples treated with NaOH were fully hydrolyzed. The produced scaffolds were then thermally treated (130 °C, 2 h) and immersed in methanol for coagulation for 1 h. The results indicate high resistance of the protein scaffolds to immersion in PBS buffer for 28 days. These scaffolds were then used for the culture of adipose mesenchymal cells. The results also suggest that the thermal treatment and coagulation could have cross-linked the fibers and help in the resistance to water. There was a significant adhesion and differentiation of adipose mesenchymal stem cells in the 3D protein scaffolds [14].

Similarly, blends of SPI-PEO (1:4, 2:3, and 3:2) were used to prepare electrospun nanofibers and tested for wound dressing applications [23]. SPI in dilute NaOH and PEO (300 kDa) in ethanol were prepared separately, then blended at different ratios of SPI:PEO before electrospinning. The results indicate that the blend with higher content of SPI (3:2) of SPI did not produce fibers but beads. In addition, it was observed that fibers collected on rotary drum resulted in higher diameter than those from stationary collector. The electrospun SPI/PEO fibers showed a lower moisture vapor transmission rate, higher tensile strength, and lower elongation at break compared to electrospun fibers of PEO alone. The SPI/PEO mats exhibit antibacterial activity against one gram-negative (*Pseudomonas aeruginosa*) and one gram-positive bacteria (*Staphylococcus aureus*) as determined by the disk diffusion method. Although the experiments for wound healing effect were qualitative observations on winstar rats (i.e., no quantification of ephitelial cell growth), the comparison indicates that the wounds covered with SPI/PEO electrospun mats exhibit a slightly better healing ability compared to uncovered wound [23].

Recent efforts in wound dressing include the use of SPI-PEO scaffolds produced by electrospinning for wound healing on a porcine model using a full thickness scission wound [11]. Soy protein isolate (7 wt%) with 0.5 % w/v PEO (1000 kDa) was dissolved in 1,1,1,3,3,3-hexafluoro-2-propanol for eight days prior to electrospinning. The produced soy protein scaffolds (3 cm × 3 cm) were applied to the wounds at different times. The results indicate that the protein scaffolds in contact with the skin get completely hydrated and dissolved in the wound. The histological analysis of the wounds revealed the positive effect of soy protein scaffolds on the wound healing process, including decrease in the number of inflammatory cells, enhancement of collagen formation in the wounded skin (dermal tissue regeneration, see Figure 6.4), regeneration of critical dermal appendages

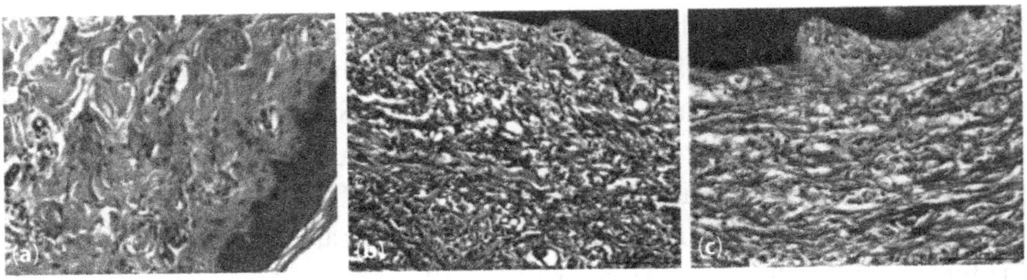

Figure 6.4 Micrograph of tissue from porcine model after 4 weeks injury, stained with Masson's trichrome (MTS). (a) Healthy skin taken close to the untreated control. (b) Wound area of an untreated control. (c) Wound area of soy protein scaffold treated wound. Purple stained areas correspond to collagen [11]. Reprinted from Wound Medicine, Vol. 5, Yah-el Har-el, Jonathan A. Gerstenhaber, Ross Brodsky, Richard B. Huneke, Peter I. Lelkes. Electrospun soy protein scaffolds as wound dressings: Enhanced re-epithelialization in a porcine model of wound healing. Pages 9–15., Copyright © 2014 with permission from Elsevier.

(hair, sweat glands), and re-ephitelialization right after 2 weeks of treatment. It was hypothesized that the healing mechanism could be due to proteolytic degradation of the soy scaffolds, which liberates the peptides in soy protein [11].

The effect of different processing parameters on the production of SPI/PEO electrospun fibers for tissue scaffolds was also recently studied [24]. Different variables included the content and molecular weight of PEO (100 kDa, 1 MDa, and 8 MDa), and different ratios of SPI/PEO in precursor solutions (SPI (7%)-PEO (3%) and 12% SPI/3% PEO). The polymers were dissolved in alkaline media (1% NaOH) and thermally treated (60 °C, pH 13, 2 h) [24]. Untreated and cross-linked (10 mM EDC/4 mM NHS, 24 h) soy protein nanofiber mats were compared. The increase in PEO molecular weight and concentration produced an increase in fiber diameter. In addition, fiber diameter was increased with the applied voltage (15–27 kV), with 27 kV producing thicker and more uniform fibers. For a constant concentration of PEO, the increase of SPI concentration produced an increase in viscosity of solutions and a decrease in surface tension, which in turn produced an increase in fiber diameter. Cross-linking of the fiber mats produced an increase in fiber diameter and the joining of the fibers at their intersection points. For a constant content of PEO, the Young's modulus increased (110–170 kPa) with the content of SPI on the blend (7–12%); an effect attributed to the increase in fiber diameter, the opposite effect was observed when blends of SPI PVA were used. The produced scaffolds were then used for human mesenchymal stem cell (hMSC) culture. Successful cell growth and adhesion were observed on the cross-linked scaffolds, with higher growth observed for the fibrous mat with 12% SPI [24].

6.3.2 Filtration and Antibacterial Applications

Soy flour was purified by isoelectric precipitation and thermal treatment at 60 °C, obtaining two protein fractions with protein content higher than the original soy flour. These fractions were compared with soy protein isolate and soy protein concentrate on its ability to produce nanofibers by electrospinning of an alkaline solution (pH 11) and using an SPI/PEO (600 kDa) ratio of 7:3. The electrospinning was carried out on a needle-free device, in which the jets originate at the surface of the polymer solution. This allowed for a faster production rate. All the solutions were thermally treated (60 °C, 30 min) before electrospinning. The results indicate that polyethylene oxide PEO enhanced the spinnability of the soy protein solutions; however, concentrations of PEO lower than 30 wt% (on a dry basis) did not produce uniform fibers. To test the bacterial efficiency of these nanofibers against gram negative bacteria *Escherichia coli* K-12, the nanofibers were electrospun on the surface of filter paper and a nonwoven fabric, which served as support layers. The bacteria solution was then aerosolized and forced to pass through the filtration media by applying vacuum; then, the flow exiting from the filter was put into contact with an agar plate, which allowed counting of bacteria colonies. The results indicate that filter media covered with electrospun soy/PEO nanofibers exhibited a high bacterial filtration efficiency compared to neat filter media. In addition, the bacterial filtration efficiency increased with the load of nanofibers, with 5 g/m^2 exhibiting the highest efficiency [12]. It was hypothesized that the adhesion of bacteria to the nanofibers might be as a result of electrostatic interactions, owed to the charge balance of aminoacids on the protein [12].This hypothesis, however, would imply that the nanofibers are positively charged, which is unlikely given the high amount of negatively charged aminoacids (mostly aspartic and glutamic acid) on soy protein and the alkaline pH of precursor solutions. The adhesion mechanism could be due to interception of the bacteria on account of the high surface area. The results did not indicate the value for fiber diameter (which in turn affects the surface area) of the produced nanofibers and their efficiency in bacteria filtration.

Similar adhesive patches of soy protein nanofibers supported on rayon fibers have been proposed as filtration membranes against esca fungi [22]. Two commercial soy protein isolates were used, one with enhanced water solubility to produce transparent solutions and other without any modification. Solutions of water soluble soy isolate were prepared in water/alcohol and mixed with PVA. On the other hand, the solutions of insoluble soy isolate were prepared using a mixture of formic acid and acetic acid and mixed with PCL. The fibers produced from the soluble isolate/PVA produced uniform fibers with an average diameter of 710–800 nm regardless of the molecular weight of the PVA used, however, the fibers prepared from insoluble isolate/PCL were thicker (~1·1.2 μm). When the adhesive properties of the fibers were tested on balsa wood substrates, the results indicated an increase in peeling force, specific and normal adhesive force with the content of adhesive. The exposure of these fiber mats to atmospheric conditions during 30 days produced two outcomes: the fibers prepared from soluble isolate/PVA did not resist water, but those prepared from insoluble soy isolate/PCL did not dissolve [22].

Other applications include the encapsulation of molecules with specific functionalities, such as natural antibacterial agents and antioxidants. For instance, Vega-Lugo and Lim [16] used SPI/PEO electrospun fibers to encapsulate up to 20 wt% of allyl isothiocyanate (AITC). The AITC was incorporated in the electrospining solutions alone and also with β-cyclodextrin (incorporation of AITC in cyclodextrin prior to mixing with the SPI solution). To help with solubility and encapsulation, it was necessary that the use of ultrasound for short time (1 min) to produce uniform fibers. Otherwise, beads or electrospraying was observed. The addition of AITC at concentrations higher than 3% produced fibers with smaller diameter than those obtained without AITC, these fibers also exhibit considerable bead formation, a change that was attributed to the interactions of the AITC with the surfactant. The release of AITC from the produced fibers increased with the exposure to higher relative humidity (up to 75% RH).In addition, the release was enhanced for those fibers produced from solutions containing β-cyclodextrin. This behavior suggests that the material could serve as an alternative for food packaging applications.

The encapsulation of antioxidants onto nanofibers has been proposed for food preservation and antibacterial applications. Natural antioxidants (anthocyanins from raspberry extracts) were added (10%) to SPI solutions with concentrations up to 10 wt% and 0.5 wt% PEO dissolved in 1 wt% sodium hydroxide to prepare electrospun fibers [25]. The extracts were added either after or before thermal treatment (heating up to 60 °C, 2 h). Although the system exhibit fiber formation ability, there was a significant amount of bead formation, which was not affected by the presence of raspberry extracts. However, there were no attempts in the study to optimize fiber forming conditions. The total amount of anthocyanins decreased (to about half) due to the thermal treatment of the samples [25].

6.4 Conclusion and Outlook

Soy proteins have proven to be the interesting biopolymers for a variety of applications, including electrospinning. Soy protein can be electrospun into fibers depending on the physicochemical conditions of the protein solutions. Different studies of the right conditions to produce fibers indicate that the use of coadjutant polymers can help in the electrospinning process. In some cases, blending soy protein with other biopolymers also helps in the production of nanofibers by electrospinning. However, the blends with other polymers are somehow restricted to the use of water-soluble polymers, which also affects the physical properties of the final electrospun fibers.

Different applications have been proposed for the soy protein electrospun nanofibers, from encapsulation of active components, filtration, and wound dressing. However, some challenges for practical applications still remain, such as the improvement of mechanical properties and water resistance of the electrospun soy nanofibers. The use of cross-linking agents has proven helpful to solve the latter, but there is a need to find nontoxic cross-linkers. In addition, applications in wound dressing would require additional testing for allergenicity and adverse reactions to the soy protein scaffolds.

Soy protein alone or in combination with other polymers produced by electrospinning represents an economic and feasible alternative to produce nanofibers by electrospinning with potential for using in different applications.

References

1. Murphy, P.A.,In *Soybeans: chemistry, production processing, and utilization*, L.A. Johnson, P.J. White, R. Galloway, Eds., AOCS press, Urbana, IL, pp. 229–267, 2008.
2. Kinsella, J., *J. Am. Oil Chem. Soc.*, 56, 242–258, 1979.
3. Peng, I.C., Quass, D.W., Dayton,W.R., Allen, C.E., *Cereal Chem.*, 61, 480–490, 1984.
4. Koshiyama, I., *Cereal Chem.*, 45, 394–404, 1968.
5. Koshiyama, I., Fukushima, D., *Cereal Chem.*, 50, 114–121, 1973.
6. Lin, L. (2011). Electrospun Soy Protein-based Scaffolds for Skin Tissue Engineering and Wound Healing (Doctoral Dissertation, Drexel University). Retrieved from https://idea.library.drexel.edu/islandora/object/idea:3437

7. Hansen, C., *Hansen Solubility Parameters: User's Handbook*, 2nd edn., CRC Press, Boca Raton, FL, USA, 2007.
8. Aghanouri, A., Sun, G., *RSC Adv.*, 5, 1890–1892, 2015.
9. Xu, X., Jiang, L., Zhou, Z., Wu, X., Wang, Y., *ACS Appl. Mater. Interf.*, 4, 4331–4337, 2012.
10. Lin, L., Perets, A., Har-el, Y.-e., Varma, D., Li, M., Lazarovici, P., Woerdeman, D.L., Lelkes, P.I., *J. Tissue Eng. Regener. Med.*, 7, 994–1008, 2013.
11. Har-el, Y.-e., Gerstenhaber, J.A., Brodsky, R., Huneke, R.B., Lelkes, P.I., *Wound Med.*, 5, 9–15, 2014.
12. Lubasova, D.,Netravali, A., Parker, P., Ingel, B., *J. Nanosci. Nanotechnol.*, 14, 4891–4898, 2014.
13. Lubasova, D., Mullerova, J., and Netravali, A.N., *J. Appl. Polym. Sci.*, 132, 41852, 2015.
14. Xu, H.,Cai, S., Sellers, A., and Yang, Y., *RSC Adv.*, 4, 15451–15457, 2014.
15. Vega-Lugo, A.C., Lim, L.T., *J. Biobased Mater. Bioenerg.*, 2, 223–230, 2008.
16. Vega-Lugo, A.C., Lim, L.T., *Food Res. Int.*, 42, 933–940, 2009.
17. Cho, D., Netravali, A.N., Joo, Y.L., *Polym. Degrad. Stab.*, 97, 747–754, 2012.
18. Cho, D., Nnadi, O., Netravali, A., and Joo, Y.L., *Macromol. Mater. Eng.*, 295, 763–773, 2010.
19. Wang, H., Jiang, Q., Jiang, S., and Jiang, S., *Polym. Polym. Compos.*, 20, 621–627, 2012.
20. Phiriyawirut, M., Rodchanacheewa, N., Nensiri, N., Supaphol, P., *Adv. Mater. Res.*, 2008.
21. Salas, C., Ago, M., Lucia, L.A., Rojas, O.J., *React. Funct. Polym.*, 85, 221–227, 2014.
22. Sett, S., Lee, M.W., Weith, M., Pourdeyhimi, B., Yarin, A.L., *J. Mater. Chem. B*, 3, 2147–2162, 2015.
23. Thirugnanaselvam, M., Gobi, N., Arun Karthick, S., *Fibers Polym.*, 14, 965–969, 2013.
24. Ramji, K., Shah, R.N., *J. Biomater. Appl.*, 29, 411–422, 2014.
25. Wang, S., Marcone, M.F., Barbut, S., Lim, L.T., *Food Res. Int.*, 52, 467–472, 2013.

7

Soy Proteins as Potential Source of Active Peptides of Nutraceutical Significance

Junus Salampessy and Narsimha Reddy*

School of Science and Health, Western Sydney University, Penrith, Australia

Abstract

Soy bean is a well-recognized food item that is consumed all over the world. Nutritionally, soybean proteins contain a complete range of essential amino acids that make them comparable to other proteins from animal and plant sources. Hence the reason for its importance in human diet. Abundant literature demonstrates that the soy protein-derived peptides display several physiological activities including antihypertensive, antioxidative, immunomodulating, hypocholesterolemic, and anticancer properties and hence are of great importance in the development of pharmacologically active constituents and functional foods. Therefore, these peptides offer significant prospect for application in food and pharmaceutical industry. However, their enzymatic production, purification and characterization are time-consuming and expensive processes. Also, the quantity of active peptides produced from hydrolysis of food/soy proteins depends on the number of occurrences of the desired peptide sequences within the parent proteins and the specificity of the enzymes used in hydrolysis. The yield of bioactive peptides produced by this method is usually low and more advanced methodologies/technologies are required to improve their yield. This chapter is devoted to review the literature to identify and describe the available methodologies for the identification and production of bioactive peptides from soybean proteins. The potential applications of soy protein-derived bioactive peptides as functional foods and therapeutic agents will also be emphasized in this chapter.

***Keywords*:** Soy protein-derived peptides, bioactive peptides, enzymatic hydrolysis, *In silico* hydrolysis, functional foods, therapeutic agents, recombinant DNA technology, fermentation, chemical synthesis

**Corresponding author*: n.reddy@westernsydney.edu.au

Visakh P. M and Olga Nazarenko (eds.) Soy Protein-Based Blends, Composites and Nanocomposites, (155–194) 2017 © Scrivener Publishing LLC

7.1 Introduction

Soy bean is a well-recognized food item that is consumed all over the world. Nutritionally, soybean proteins contain complete range of essential amino acids that make them comparable to other proteins from animal and plant sources. Hence the reason for its importance in human diet. Abundant literature demonstrates that the soy protein-derived peptides display several physiological activities including antihypertensive [1], antioxidative [2], immunomodulating [3], hypocholesterolemic [4], and anticancer properties [5] and hence are of great importance in the development of pharmacologically active constituents and functional foods. Therefore, these peptides offer significant prospect for application in food and pharmaceutical industry. However, their enzymatic production, purification, and characterization are time-consuming and expensive processes. Also, the quantity of active peptides produced from hydrolysis of food/soy proteins depends on the number of occurrences of the desired peptide sequences within the parent proteins and the specificity of the enzymes used in hydrolysis. The yield of bioactive peptides produced by this method is usually low and more advanced methodologies/technologies are required to improve their yield. This chapter is devoted to review the literature to identify and describe the available methodologies for such purposes. The chapter aims to cater to the needs of beginners as well as the experienced researchers in the field.

Recent developments in food bioactive protein/peptide databases, coupled with improved knowledge of various enzyme specificities can be used in a process known as *in silico* hydrolysis for the identification of potential bioactive peptides from food/soy proteins. *In silico* produced peptides with known sequences can then be subjected to quantitative structure-activity relationship (QSAR) studies for a preliminary assessment of their bioactivity potential [6]. This improved knowledge can then be used to subject the food protein to hydrolysis with identified enzymes to speed up the process of production and screening. The *in silico* hydrolysis protocols have therefore been extensively reviewed in this chapter.

Fermentation is another important process that provides ways to enhance nutritive value of foods and also helps to produce bioactive peptides. Fermentation of soybean has been shown to result in the release of peptides with various functionalities [5] and these aspects have also been reviewed extensively in this chapter. Chemical synthesis of peptides of known sequences and desired functionalities is an alternative method for the production of large quantities of highly pure bioactive peptides [7,8]. Solution- and solid-phase syntheses are the two main methods for chemical synthesis of bioactive peptides. At present, solid-phase peptide synthesis is considered as a more established method and has been used to synthesize significant numbers of bioactive peptides [7]. There are several advantages to produce bioactive peptides by chemical synthesis, which include producing peptides of high purity and quantity, desired sequences that are otherwise difficult to obtain from natural sources, and peptides of known activity identified in natural sources but are difficult to ensure their release through hydrolysis. This approach can be used to produce bioactive peptides of all physiological functionalities and has been briefly reviewed in this chapter.

The chapter then brings about the drawbacks of conventional methods of production of bioactive peptides and suggests possible ways to alleviate such shortcomings. The enzymatic hydrolysis and fermentation methods discussed above are the most commonly used techniques in preparation of bioactive peptides and both of them suffer from time consuming and expensive processes that produce low quantities of specific active peptides. On the other hand, chemical synthesis improves yield and purity but involves expensive processes. Recombinant DNA technology is an alternate and efficient method that has recently been used to produce bioactive peptides identified from food proteins [9]. This approach results in the production of proteins containing multicopies of the desired sequences of bioactive peptides. Production of recombinant proteins containing repeated sequences of active peptides can improve the quantity of bioactive peptides that can be produced by subsequent enzyme hydrolysis. At the present time, recombinant DNA technology for protein production remains in its infancy. However, current research findings indicate that the technology for the production of recombinant proteins containing encrypted bioactive peptides is feasible. The main features of recombinant DNA technology and its applications to improve the yield of bioactive peptides has been reviewed and highlighted in this chapter.

Potential applications of soy protein-derived bioactive peptides as functional foods and therapeutic agents have also been emphasized in this chapter.

7.2 Soy Proteins as Source of Bioactive Peptides

Soy bean is a well-recognized staple food in Asia and is used in many preparations either as the main source of nutrient (proteins) or to flavor other preparations [10, 11]. Products such as soy milk, tempeh, and tofu are the examples of soy products consumed as protein sources, whereas soy sauce and miso are examples of soy products used to flavor other food preparations. Soy bean contains large quantities of protein that account for about 44% (db) or 38% (wb) [10, 12]. Hence, soybeans are a very good source of proteins for human consumption. Also, it is a nutritionally superior food when compared to oilseeds and cereals [11].

Soy bean proteins, as seed proteins consist of mostly storage proteins, in addition to metabolic proteins. The storage proteins are synthesized during soybean seed development [10] and function as major protein source for human consumption. The two main types of storage proteins in soybean are glycinin and β-conglycinin, also known as 11S and 7S protein, respectively [10, 13]. The glycinin consists of five subunits, that is, G1, G2, G3, G4, and G5 subunits; the β-conglycinin consists of three subunits, that is, α, α', and β subunits [14]. However, soybean contains Bowman-Birk inhibitor (BBI) which is an inhibitor of trypsin and chymotrypsin [15, 16] that have been associated as antinutritional agent [17–21].

Nutritionally, soybean proteins contain complete amino acids that make them comparable to other proteins from animal sources [14]. These proteins contain all amino acids essential to human nutrition [10] (Table 7.1), hence the reason for its importance in human diet. Major applications of soy protein apart from its use for human consumption are in paper coatings [22] and as animal feed [10]. Currently, however, soybean proteins are used in the preparation of functional peptides such as antioxidative [23–25], antihypertensive [26–28], and anticancer [23, 29] peptides. These physiological activities of peptides are of great importance and have relevance in

Table 7.1 Amino acid composition of soybean proteins*.

Non-essential amino acids	Content (%)	Essential amino acids	Content (%)
Alanine	1.685	Arginine	2.706
Aspartic acid	4.469	Histidine	0.968
Glutamic acid	6.362	Isoleucine	1.606
Glycine	1.660	Leucine	2.838
Cysteine	0.407	Lysine	2.310
Proline	1.948	Methionine	0.405
Serine	2.028	Phenylalanine	1.926
Tyrosine	1.341	Threonine	1.315
		Tryptophan	0.49
		Valine	1.662

*Expressed in gr/100gr seed or %.
Adopted from Goldflus *et al.* [30].

the development of pharmacologically active constituents and functional foods. In this review, emphasis is given to identify and list physiologically active peptides that can be derived from soybean proteins together with their production methods and potential application as therapeutic agents as well as nutraceuticals for functional food preparation.

7.3 Identification of Potential Bioactive Peptides from Soy Proteins

Physiologically active peptides derived from food proteins have gained a major research interest in the last few decades. Their production, identification/isolation and characterization have been based mostly on functional-guided screening followed by time-consuming purification procedures and chemometric-guided characterization [24, 31–34]. However, as time progressed, data obtained from numerous research activities in this field have been reported and databased which can be accessed by researchers from around the world. Databases such as BIOPEP [35] and Antimicrobial Peptide Database (http://aps.unmc.edu/AP/main.php) are two such examples of these databases. On the other hand, the advent of sophisticated proteomic analysis and advanced mass spectrometric methods has led to the analysis of primary sequences of a great numbers of proteins and peptides. These advancements, coupled with improved knowledge of various enzyme specificities can be used in a process known as *in silico* hydrolysis for the identification of potential bioactive peptides from proteins with known primary structures.

Various bioactive peptides have been identified from soybean proteins [6] (Figure 7.1). As can be seen from the figure, soybean proteins are good source of small bioactive peptides as several known sequences are matching with soybean protein, glycinin G1 in this example. However, it is unlikely that hydrolysis or fermentation of soybean glycinin G1 will produce all these active peptides as this production depends on the specificity of enzymes and microbes used for the hydrolysis or fermentation. Therefore, the use of database such as BIOPEP and Antimicrobial Peptide Database are useful mainly for identification purpose. However, the use of these databases in combination with other *in silico* hydrolysis of proteins of known sequences can provide plausible information on the potential bioactive peptides that can be produced and the suitable enzymes to hydrolyse the proteins. Produced peptides with known sequences but without known activities can then be subjected for quantitative structure-activity relationship (QSAR) to predict their potential.

```
        10         20         30         40         50         60
MAKLVFSLCF LLFSGCCFAF SSREQPQQNE CQIQKLNALK PGNRIESEGG LIETWNPNNK
        70         80         90        100        110        120
PFQCAGVALS RCTLNRNALR RPSYTNGPQE IYIQQGKGIF GMIYPGCSST FEEPQQPQQR
       130        140        150        160        170        180
GQSSRPQDRH QKIYNSREGD LIAVPTGVAW WMYNNEDTPV VAVSIIDTNS LENQLDQMPR
       190        200        210        220        230        240
RFYLAGNQEQ EFLKYQQEQG GHQSQKGKHQ QEEENEGGSI LSGFTLEFLE HAFSVDKQIA
       250        260        270        280        290        300
KNLQGENEGE DKGAIVTVKG GLSVIKPPTD EQQQRPQEEE EEEDEKPQC KGKDKHCQRP
       310        320        330        340        350        360
RGSQSKSRRN GIDETICTMR LRHNIGQTSS PDIYNPQAGS VTTATSLDFP ALSWLRLSAG
       370        380        390        400        410        420
FGSLRKNAMF VPHYNLNANS IIYALNGRAL IQVVNCNGER VFDGELQEGR VLIVPQNFVV
       430        440        450        460        470        480
AARSQSDNFE YVSFKTNDTP MIGTLAGANS LLNALPEEVI QHTFNLKSQQ ARQIKNNNPF
       490
KFLVPPQESQ KRAVA
```

Figure 7.1 Example of identification of bioactive peptides from glycinin G1 protein. Angiotensin I converting enzyme (ACE)-inhibitory peptides (ACEIPs; Shade with dotted underline), antioxidative peptides (AOPs; Dotted underline), DPP-IV inhibitors (DPP-IVPs; Shade), peptides that function as both ACEIP and AOP (Underline), peptides that function as both ACEIPs and DPP-IVPs (Shade with underline) and peptides that function as both AOPs and DPP-IVPs (Double underline).

A combined use of *in silico* hydrolysis and peptide database-guided identification is illustrated in Figure 7.2. The enzymes used for this illustration are Asp-N endopeptidase, pepsin, trypsin, and chymotrypsin. These enzymes are commercial enzyme and the main digestive enzymes in animals and humans. Pepsin cleaves proteins to produce peptides containing amino acids with aromatic side chains at both N- and C-termini and also the peptides containing leucine, aspartic acid, and glutamic acid at C-termini. Trypsin and chymotrypsin cleave proteins to produce peptides containing lysine, arginine, phenylalanine, tyrosine, and tryptophan at C-termini [36]. As can be seen from the figure, much less small peptides are produced and of these peptides only few are known active peptides. The activity of the rest is yet to be determined either through synthetic production and activity analysis, through QSAR studies or combination of both approaches. It is also noteworthy that bioactive peptides can only be released if a suitable enzyme or a combination of enzymes is employed. Therefore, *in silico* digestion offers a good approach to gain insight into the suitable enzymes that can be employed to produce bioactive peptides. One of the limitations of using *in silico* digestion approach is the limited number of enzymes that are available for this purpose. Table 7.2 presents the list of active peptides derived from *in silico* hydrolysis of soybean proteins

```
        10         20         30         40         50         60
MAKLVFSLCF LLFSGCCFAF SSREQPQQNE CQIQKLNALK PGNRIESEGG LIETWNPNNK
        70         80         90        100        110        120
PFQCAGVALS RCTLNRNALR RPSYTNGPQE IYIQQGKGIF GMIYPGCSST FEEPQQPQQR
       130        140        150        160        170        180
GQSSRPQDRH QKIYNSREGD LIAVPTGVAW WMYNNEDTPV VAVSIIDTNS LENQLDQMPR
       190        200        210        220        230        240
RFYLAGNQEQ EFLKYQQEQG GHQSQKGKHQ QEEENEGGSI LSGFTLEFLE HAFSVDKQIA
       250        260        270        280        290        300
KNLQGENEGE DKGAIVTVKG GLSVIKPPTD EQQQRPQEEE EEEDEKPQC KGKDKHCQRP
       310        320        330        340        350        360
RGSQSKSRRN GIDETICTMR LRHNIGQTSS PDIYNPQAGS VTTATSLDFP ALSWLRLSAG
       370        380        390        400        410        420
FGSLRKNAMF VPHYNLNANS IIYALNGRAL IQVVNCNGER VFDGELQEGR VLIVPQNFVV
       430        440        450        460        470        480
AARSQSDNFE YVSFKTNDTP MIGTLAGANS LLNALPEEVI QHTFNLKSQQ ARQIKNNNPF
       490
KFLVPPQESQ KRAVA
```

Figure 7.2 Example of *in silico* hydrolysis with Asp-N in combination with GI enzymes (pepsin, trypsin and chymotrypsin) and identification of bioactive peptides from glycinin G1 protein. Angiotensin I converting enzyme (ACE)-inhibitory peptides (ACEIPs; Dotted underline), DPP-IV inhibitors (DPP-IVPs; Double underline) and peptides that function as both ACEIPs and DPP-IVPs (Wave underline). Predicted ACEIPs are in bold, predicted AOPs are italic and predicted DPP-IVPs are underlined. Peptide sequences represented with shade are not assigned.

with same sequences as those active peptides reported from other sources of proteins. Table 7.3, on the other hand, presents the list of predicted bioactive peptides derived through *in silico* hydrolysis of soy bean storage proteins and suitable enzymes available for this hydrolysis.

As can be seen from Tables 7.2 and 7.3, numerous bioactive peptides can be derived from various soybean storage proteins [6]. The enzymes used for this purpose are pepsin, high specificity chymotrypsin, trypsin, Arg-C proteinase (clostripain; clostridiopeptidase B), Asp-N endopeptidase, thermolysin and combination of these individual enzymes with gastrointestinal enzymes pepsin, chymotrypsin and trypsin. Pepsin (EC 3.4.23.1), chymotrypsin (EC 3.4.21.1), and trypsin (EC 3.4.21.4) are gastrointestinal enzymes commonly found in vertebrates and are instrumental in human digestion of consumed proteins. Clostripain (EC 3.4.22.8) also known as Arg-C peptidase [38] is a sulfhydryl protease produced by *Clostridium histolyticum* and prefers arginine over lysine residue. Under controlled hydrolysis conditions it will cleave proteins at peptide bonds limited to arginine P1 sites, including trypsin-resistant arginylprolyl bonds [39], and produce peptides containing arginine residues at C1 positions (C-terminal position 1). Asp-N endopeptidase (EC 3.4.24.33) is a metallopeptidase isolated

Table 7.2 Identified bioactive peptides derived from soybean storage proteins through *in silico* hydrolysis with gastrointestinal enzymes and some commercial enzymes[*,a,b].

Proteins	Activity		
	ACE inhibitors	**Antioxidants**	**DPP IV inhibitors**
β-conglycinins	RL, IR, IY, VF, PR, LF, GP, PL, FFL, AF, RA, VG, IG, GL, GR, FG, DA, GK, QG, LG, QK, NF, NK, AR, EI, IE, LN, EK, PH, HK, AIP, GPL, LKP	IR, LK, IY, AY, LLPH, PHF, RHK, LKP	AF, AS, AY, DN, DP, DR, EK, ES, ET, GA, GP, IR, LL, MR, NP, NQ, PH, PL, QG, QN, QT, RH, SH, SK, VE, VG, VN, VQ
Glycinins	IR, PR, GP, AR, IY, GR, IF, GK, GS	IR, LK, ADF	AF, AG, AH, DN, DQ, DR, EG, EH, GG, GP, HE, HL, IQ, IR, MG, MK, MQ, MR, NA, NG, NL, NN, NR, QN, QQ, SH, SK, SV, TH, TL, TN, VA, VF, VH, VK, VQ, VS, VT
Protease inhibitors	VG, IG, GK, VK	PEL	DR, LV, VK,

*Only active peptides identified and reported from other food proteins are presented here.
[a]Only enzymes that are commonly used for protein hydrolysis are used here.
[b]The activities of these peptides can be obtained from BIOPEP Database [37].

Table 7.3 Predicted bioactive peptides derived from soybean storage proteins through *in silico* hydrolysis with gastrointestinal enzymes and some commercial enzymes*[,a,b].

Proteins	Activity		
	ACE inhibitors*	Antioxidants[b]	DPP IV inhibitors[b]
β-conglycinins[c]	LE, LL, LK, *MR*, *VN*, *IN*, NQY, LQR, **IVI**, **PHF**, AEL, *LDI*, *VIP*, *AGY*, *VQE*, GSNR, **EQIR**, SSSR, LEVR, SPQL, EEQEW, **TISSE**, VINEG, DANIE, *VEISK*, *VPSGTT*, *VNKPGR*, NSKPNT, DMNEGA, **VIPAGY**, **GPLSSI**, **NQACHAR**	RH, PLL, GPL, AGY, DAQPK, VIPAGY, VIPAAY, EEPQQK, SPQL, NPQL, ASVSVS, EEQW	VIPAGY, VIP. AIP, GIA, ENPK, NPF, NPQL, EEK, GEK, EVR, LEVR, NQY, GPL, GPLSSI, AIVI
Glycinins[c]	VR, LK, IK, *MR*, IET, **GIF**, **GGL**, **EAF**, **IQT**, **VVA**, **GAL**, **QQL**, ***VTR***, **DQL**, **DVF**, **LEP**, **PAL**, **GQL**, *IQK*, *VHR*, *IQR*, *ADF*, *VTS*, *VNP*, VPH, PSY, SGCC, **GMIY**, QEGR, **GMIF**, **LVNP**, **GCQR**, **QIVR**, **SSAC**, **QQAR**, SQQQ, **DEEE**, *IDET*, *PEEV*, *ESQR*, *VQGK*, *VEGG*, VEEN, QSQK, SVDK, RPSY, VVDR, ESQR, SPHW, IVPQN, GSQSK, **VVAAR**, **SQQAR**, **SQRPQ**, **AVAAK**, **AVAAR**, **SPYPR**, SHGGR, GCQTR, **VVPQN**, *IQQGK*, *AVPTG*, *SQSDN*, *SPYPQ*, *ARPSR*, *ITSSK*, IESEGG, **KPQQEE**, **PGCPST**, PGCPET, LIPSEV, *PGCSST*, *VESEGG*, *RPSHGK*, *VNCQGN*, EEPQQK, IAVPTG, AITSSK, SVISPK, RPSHGK, QGNSGP TNGPQEI, GQSSRPQ, AGNQEQE, **GAIVTVK**, SVIKPPT, DEKPQCK, VTAPAMR, DRPSIGN, LQGENEGE, DETICTMR, **PEEVIQHT**, GVNMQIVR, LQGENEEE, **PEEVIQQT**, CAGVTVSK, QIVTVEGG, *AQQNECQ*, *SQRPQDR*, *IGQTSSPDI*	ENEGE, EGEDK, AGNQEQE	VVA, AVA, MAK, LAQS, APEF, VTAP, PAL, VVA, NPK, NPQ, NPIY, AEK, DEK, NAL, PAL

(*Continued*)

Table 7.3 Cont.

Proteins	Activity		
	ACE inhibitors*	**Antioxidants**[b]	**DPP IV inhibitors**[b]
Protease Inhibitor	*VK*, SSW, PEL, **ENP**, *IER*, *ITD*, *VQGT*, **DGIVK**, *SDDCK*, **EVPHIG**, *IVPEGN*, MSDDCK, EVPHIG		EVPH, ENP,

[*] The activities of these ACE inhibitory peptides have been predicted using QSAR approach [6]
[a] Only enzymes that are commonly used for protein hydrolysis are used here.
[b] Activities are predicted based on the presence of known active sequences.
[c] Listed peptides are the common peptides derived from each member of the β-conglycinins and the glycinins.
Bold: released by Asp-N+GI enzymes only
Italics: released by thermolysin+GI enzymes only
Underlined: released by GI enzymes only.
Unmarked sequences are released by all enzyme combinations. Sequences with double marking are released by the respective enzyme combination.

from *Pseudomonas fragi* with a narrow specificity similar to flavastacin (EC 3.4.24.76) and cleaves protein at peptide bonds mostly to aspartic acid (D) at P1′ positions and, to a slower rate, at glutamic and cysteic acid bonds [40] hence produce peptides containing mainly aspartic acid (D) at N1 positions (N-terminal position 1). Thermolysin (EC 3.4.24.27) is a metalloendopeptidase that preferentially cleaves proteins and produces peptides with bulky and aromatic residues (I, L, V, A, M and F) at P1′ positions [41] hence will produce peptides containing these amino acid residues at N1 positions.

In silico hydrolyses of β-conglycinins (α, α', and β subunits) have shown that among the enzymes employed thermolysin is the most effective single enzyme while Asp-N endopeptidase is the least effective single enzyme [6]. However, in combination with gastrointestinal enzymes (pepsin, chymotrypsin, and trypsin), Asp-N endopeptidase produces the greatest numbers of peptides when compared to the numbers of peptides produced by other enzyme combinations [6]. Many peptides produced from these hydrolyses have also been produced from other food protein sources and reported extensively in scientific literature to have ACE-inhibitory activity (Table 7.2). These peptides include, among others, RL, IR, IY, VF, PR, LF, FY, GP, PL, AF, VG, IG, GL, GR, FG, DA, GK, QG, LG, QK, NF, NK, AR, IE, LN, EK, PH, HK, AIP, and GPL [38]. In addition, there are also peptides that contain known sequences of ACE-inhibitory dipeptides and tripeptides. These peptides and some other unknown peptides have been subjected for QSAR analyses and the predicted active peptides are presented in Table 7.3 [6]. Since the β-conglycinins share 80% sequence identity within the group [42], a great numbers of peptides with identical sequences are produced from the hydrolysis of these proteins. For example, the peptides LE, LL, LK, NQY, LQR, GSNR, SSSR, LEVR, EEQEW, VINEG, DANIE, and NSKPNT, having predicted ACE-inhibitory activities, IC_{50} values, of 87.4, 61.9, 44.0, 49.1, 39.0, 22.8, 11.8, 25.9, 8.8, 13.0, 29.9, and 28.0 μM, respectively (Table 7.3) [6]. This indicates the richness of the hydrolysates produced in term of the numbers of active peptides presence in them and the potential efficacy these hydrolysates possess. Of particular interest are IVI, PHF, VIP, LEVR, and SSSR that can be used as leads in the production of antihypertensive agents as they contain amino acid residues that can improve effective binding to the active site of ACE [43].

In silico hydrolysis of the glycinin group of proteins also produce a significantly great numbers of active peptides [6]. As mentioned earlier, this group consist of previously known five subunits (G1, G2, G3, G4, and G5 subunits) in addition to a new subunit known as G7 [44]. Various ACEIPs that have been previously identified and reported from other protein sources are produced from this *in silico* hydrolysis. These ACEIPs include IR, PR, GP, AR, IY, and GR [6, 37] (Table 7.2). Other peptides that contain known amino acid residues with strong affinities toward ACE active site as well as those that contain kwon ACEIP sequences especially those di- and tripeptides that are produced and have their ACE inhibition activities predicted [6]. These peptides include VR, IF, MR, GGL, VPH, ADF, EAF, GQL, SGCC, GMIF, PQDR, IDET, PEEV, TVPQN, SQSDN, and VVPQN and have predicted IC_{50} values of 43.08 μM, 15.72 μM, 38.79 μM, 47.29 μM, 39.28 μM, 13.66 μM, 37.95 μM, 11.73 μM, 23.24 μM, 40.41 μM, 34.83 μM, 27.72 μM, 25.66 μM, 3.59 μM, 7.62 μM, and 1.82 μM, respectively (Table 7.3). Similar to peptides produced from β-conglycinins hydrolysis, various peptides with identical sequences are also produced from the glycinins hydrolysis indicating that these are good sources of proteins as precursors of ACEIPs.

Other protein of interest in this hydrolysis is the protease inhibitor-like protein. This type of protein has been reported to possess antinutritive action and hydrolysis of this protein has been reported to degrade protease inhibitor in soybean [45]. This degradation has twofold impact, that is, increase the bioavailability of soybean protein and decreasing anti-cancer and anti-inflammatory effect of this soybean protein that has been previously reported [46]. Several ACEIPs have been identified form *in silico* digest of protease inhibitor-like protein [6] that include dipeptides VG, IG, GK, and VK that have been previously reported in the literature [37]. Other potential ACEIPs derived from this protein include SSW, ENP, IER, ITD, VQGT, and IVPEGN, having predicted IC_{50} values of 9.68 μM, 44.84 μM, 42.17 μM, 21.79 μM, 45.91 μM, and 0.28 μM, respectively. Direct consumption of intact protease inhibitors from food sources may result in indigestion of other proteins as these enzymes will inhibit the activity of gastrointestinal proteases such as chymotrypsin and trypsin. Hence hydrolysis of these enzymes prior to consumption may alleviate this problem in addition to the release of beneficial ACEIPs.

ACE-inhibitory peptides derived from food proteins during processing such as fermentation or hydrolysis are of great importance to human health. Various studies have reported the antihypertensive effects of ACE-inhibitory peptides derived from food proteins [47–50]. It is also indicated that small ACE-inhibitory peptides are less susceptible to gastrointestinal enzyme degradation and can be absorbed intact into the blood circulation [51, 52] and deliver their hypotensive effect and hence promote human health. Therefore, hydrolysis of food proteins using a combination of commercial and gastrointestinal enzymes is an effective way to produce small and active ACEIPs that can resist further gastrointestinal enzyme degradation *in vivo* and deliver their health benefits.

Antioxidative peptides are another group of active peptides that have attracted interest from researchers around the world due to their beneficial role in human health. It is therefore beneficial to identify the released antioxidative peptides from in silico hydrolysis of soybean proteins. Several antioxidative peptides derived from the hydrolysis of β-conglycinins have been shown to possess sequences similar to antioxidative peptides derived from other protein sources [6]. These peptides include IY, AY, IR, LK, PHF, RHK, LKP, and LLPH [37]. In addition, there are also a great numbers of peptides that contain sequences of known antioxidative peptides. These peptides include RH, PLL, VNPH, PHHA, SPQL, NPQL, EEQW, PHHADAD, GYALPHA, ENEGE, VDAQPK, VIPAGY, and VDAQPQQK. Various amino acid residues have been claimed to contribute to the activity of antioxidative peptides. *In silico* hydrolysis of soybean glycinins have also shown to produce antioxidative peptides (Tables 7.2 and 7.3) [6]. Various antioxidative peptides such as IR and LK [6, 37] previously reported from other food proteins have been identified from the *in silico* hydrolysis of glycinins. In addition, various potential antioxidative peptides have been identified from this hydrolysis (Table 7.3). These include VPH, SPHW, DAQPK, ENEGE, EGEDK, EEPQQK, and AGNQEQE. As major storage proteins in soybean, glycinins form a great source of potential bioactive peptides such as antioxidative peptides.

Similar to the β-conglycinins and the glycinins, protease inhibitor in soybean is also likely to release antioxidative peptides upon hydrolysis. Tripeptide PEL for example has been released through simulated hydrolysis with mixed enzymes. Other peptides with potential antioxidative activity include EVPH and EVPHIG. These peptides contain dipeptide sequence (PH) that has been part of many antioxidative peptides [37].

The storage proteins and the protease inhibitor have shown to possess sequences that can exert antioxidative activity when released from the parent proteins. Chen *et al.* (1995) [53] and Saito *et al.* (2003) [54], for example, identified antioxidative peptides from soybean proteins having 5–16 amino acid residues. These peptides consisted of hydrophobic amino acids, valine (V) or leucine (L), at the N-terminal positions, and proline (P), histidine (H) and tyrosine (Y) in any position in the sequences. These findings are in agreement with the antioxidative peptides derived *in silico* from soybean proteins as listed in Table 7.3 that consist of 2–7 residues in their sequences. Therefore, the mixed enzymatic hydrolysis simulated in the study [6] can be applied for *in vitro* hydrolysis to produce hydrolysates with antioxidative properties.

Another important activity displayed by peptides is the dipeptidyl peptidase IV (DPP-IV) inhibitory activity. This group of peptides has been the focus of research with a view to utilize them in the management of type-2 diabetes [55–58]. *In silico* hydrolysis of soy proteins with mixed commercial enzymes and gastrointestinal enzymes has also shown to produce peptides with known sequences of DPP-IV inhibitors. A numbers of dipeptides such as GP, LL, RH, GA, NP, EK, and PL can be produced from *in silico* hydrolysis of β-conglycinins. Many other tripeptides, tetrapeptides, and oligopeptides contain sequence of known DPP-IV inhibitors produced by *in silico* hydrolysis of β-conglycinins. Peptides such as GIA, NPF, NPQL, and GPFPSI are examples of *in silico* products of soy protein hydrolysis that may possess DPP-IV inhibitor activity due to their content of known DPP-IV inhibitor sequences.

In silico hydrolysis of the glycinins is also able to produce peptides of known DPP-IV inhibitory activities. Examples of these peptides include VA, AH, GP, HL, IQ, IR, MG, SV, TN, and VS which can be produced from *in silico* hydrolysis of the glycinins. Other produced peptides such as VVA, AVA, MAK, LAQS, NPQ, APEF, and VTAP may also possess DPP-IV inhibitory activity due to the fact that they contain known DPP-IV inhibitor sequences.

Likewise, *in silico* hydrolysis of soybean protease inhibitor has resulted in the production of known DPP-IV inhibitors, that is, DR, LV, and VK. Other peptides that may possess DPP-IV inhibitory activity include EVPH and ENP due to their content of known DPP-IV inhibitor sequences.

Although the mode of action of DPP-IV inhibitor peptides remains unclear, researchers have tried to reveal the relationship between structure and activity. Nongonierma and FitzGerald (2016), for example, have linked their activity to specific peptide motifs, consisting of an N-terminal (N1) threonine (W) and/or a proline (P) at position N2 [58]. *In silico* hydrolysis of soy proteins presented here have shown that soy proteins are good source of potential DPP-IV inhibitory peptides that have potentialfor the management of type-2 diabetes. This hypothesis is based on the premise that food protein hydrolysates having *in vitro* DPP-IV inhibitory activity generally yield antidiabetic effects in animal studies [58].

As can be seen from Tables 7.2 and 7.3, *in silico* hydrolysis of soy proteins has yielded a large number of potential ACEIPs. It is therefore important to look at the structural suitability of these peptides with respect to the active site structure of ACE. The overall structure of the ACE consists of several α-helices and takes an ellipsoid shape with a groove in the centre [59]. Three α-helices (α_1, α_2, and α_3) appear to form a "lid" on top of the molecule at the active site. This "lid-like" opening to the active-site is assumed to restrict the access of large molecules to the active-site. The active-site structure ofACE [59] consists of a zinc coordination site, several hydrogen bonding subsites (containing hydrogen bonding "H" residues), a proline binding pocket and a hydrophobic pocket (with "F" residues) that has preference for aromatic amino acids like phenylalanine. According to Karman *et al.* (2016) [43] it is necessary for the potent peptide inhibitors to have the following structural requirements: (1) possess strong chelating groups such as glutamic acid (E) to efficiently coordinate with zinc II ion in the ACE active-site, (2) have several functional groups that can form strong hydrogen bonding with the residues in multiple subsites within the ACE active-site, (3) have a hydrophobic residue with an aromatic side chain such as phenylalanine (F), leucine (L), methionine (M), tyrosine (Y), and threonine (W), (4) presence of proline (P) is beneficial for effective binding, and (5) preferably be short peptides with 2–4 residues (a tripeptide may possibly be the best fit). Another residue commonly found in many C-terminal ACEIPs is arginine (R) [37] for which activity has been related to its positive charge on the guanidino group of arginine side chain [60, 61]. More recently, a molecular docking study has shown the contribution of serine (S) residue at the C-terminal tripeptide TLS that is related to its ability to form hydrogen bonds with the S1 and S2 pockets of the ACE active site. Many of the ACEIPs predicted by *in silico* methodology (Tables 7.2 and 7.3) fulfil most of these criteria and are the strong candidates for ACE inhibitors.

7.4 Production of Bioactive Peptides from Soy Proteins

7.4.1 Bioactive Peptides Derived Through Fermentation

Fermentation, one of the first techniques to preserve foods, provides ways to enhance nutritive value of foods, destroy undesirable factors, to make safer products, to improve appearance and taste of some foods, to salvage raw material that otherwise not usable for human consumption, and to reduce energy used for cooking [62]. At present, fermentation continues to play an important role in providing variety of quality foods for human and domestic animal consumption. In fermentation, raw materials are converted into products through the act of micro-organisms (bacteria, yeasts, and moulds). The production of active peptides in fermentation is achieved by the action of proteolytic enzymes released by the microorganisms involved in the process. These microorganisms may be added to the raw materials as starter cultures or are part of the microorganism ecosystem of the raw materials.

In Asia, milk, soy bean, and fish are more common fermentation substrates with a range of products such as yogurt, soy sauce/paste, and fish sauce. Fermented soy bean products such as "shoyu" and "miso" in Japan, "oncom" and "tempeh" in Indonesia, "fermented tofu" in China are well known products in respective countries [63, 64]. Various bioactive peptides derived from soybean fermented products have been identified and reported (Table 7.4) [65]. These include antihypertensive/ACE inhibitory peptides, antioxidative peptides, antimicrobial peptides, DPP-IV inhibitors, antidiabetic peptides, etc. Vallabha and Tiku (2014), for example, reported the finding of an antihypertensive pentapeptide LIVTQ (IC_{50} = 0.09 μM) derived through *Lactobacillus casei* spp. *pseudoplantarum* fermentation of soybean [66]. The pentapeptide is stable against pepsin degradation under *in silico* condition and has slow absorption rate *in vitro* [66]. In another study, an ACE inhibitory pentapeptide LVQGS (IC_{50} 43.7 μM) was isolated from *Aspergillus oryzae* fermented soybean extract [67]. These findings support the previously reported production of ACEIPs from soybean fermentation [65, 68].

Fermentation of soybean has also resulted in the release of various other functional peptides. Amadou *et al.*, for example, reported the antioxidative effect of peptides derived from *Lactobacillus plantarum* Lp6 fermented soy protein meal [65, 69]. Although, the sequence/s of the active peptide/s remains unknown, the findings do indicate the presence of antioxidative peptides released from the fermentation process of soybean proteins. Hypocholesterolemic activity of peptides derived from fermented soybean product has also been reported. Oral administration of soy proteins have been reported to improve cholesterol composition of blood serum in animal models [70–72]. Chinese fermented soybean product called douchi has also been linked to antioxidant activity [73]. All this information supports the role of fermentation for the production of bioactive peptides from soybean proteins.

The cholesterol lowering effect of *Lactobacillus delbrueckii* subsp. *delbrueckii* TUA-4408L (SNC33) fermented soymilk on rats fed with high cholesterol diet has been reported [74]. It is noteworthy that fermented soymilk used in this experiment contains very little protein. Hence, the hypocholesterolemic effect may be due to bioactive peptides and other active compounds such as isoflavones and saponin. Another physiological function of soy protein is their reported ability to act as an antidiabetic. This activity is related to their ability to improve body composition and insulin secretion especially among postmenopausal women [75]. Although there is no concrete claim that fermented soy proteins can improve body composition and insulin secretion, Kwon *et al.* demonstrated that the compositional changes in isoflavonoids and peptides that occurred during a long fermentation period of "meju," a fermented soy product, enhanced the antidiabetic effect of soybeans [76]. Fermentation of soybean has also been linked to the production of some antimicrobial peptides and antithrombotic peptides. The production of antimicrobial peptides has been so far the result of starter culture release of active peptides during fermentation [5, 77, 78]. Shon *et al.* [79] had reported that fermentation of soy bean in the preparation of soy sauce produces bioactive peptides that can function as antithrombotic agents, although the peptide responsible for this activity remains unknown.

The discussions presented above duly substantiate that fermentation of soybean offers the following beneficial outcomes in terms of the production of bioactive peptides:

- Fermentation of soybean has resulted in the release of antihypertensive/ACE inhibitory peptides and antioxidative peptides.
- Consumption of fermented soybean products has been reported to improve serum cholesterol composition on animal models and may also show similar effects on human serum cholesterol profile.
- Consumption of fermented soybean products has been linked to the improvement of body composition and insulin secretion hence offer alternative ways to treat type-2 diabetes although the role of peptides produced during fermentation remains unclear and further investigations are required.
- Starter cultures used in soybean fermentation have been reported to produce antimicrobial peptides of significant strength and prospect.

7.4.2 Bioactive Peptides Derived Through Hydrolysis

Degradation or hydrolysis of protein molecules into smaller peptides and then into amino acids, their building blocks, can be achieved through the use of enzymes and acids and bases [36]. By definition, enzymatic hydrolysis of food proteins is the hydrolysis of food proteins by the action of isolated enzymes with a view to applying the hydrolysed proteins as a food ingredient [36], and the products are called protein hydrolysates.

The enzymes used in modification of food proteins can be classified according to their origin (animal, plant or microbial), their catalytic action (endopeptidase or exopeptidase), and the nature of the catalytic site. Endopeptidases are always used in food protein hydrolysis, although a combination with exopeptidases is occasionally employed [36]. The endopeptidases can be further classified based on their catalytic site into serine proteases, cysteine proteases, metalloproteases, and aspartic proteases [80, 81], while a group of threonine peptidases has also been discovered [82]. The serine proteases are well studied having a maximum activity at alkaline pH and –OH group in the catalytic centre. The cysteine proteases are similar to serine proteases but having –SH instead of –OH group in the catalytic centre. These enzymes are sensitive toward oxygen. The metalloproteases contain a metal atom, usually zinc and have optimum pH around neutral. The aspartic proteases are characterized by having a carboxyl group from aspartic acid in the active centre and maximum activity at acid pH [36].

As can be seen in Table 7.4 various bioactive peptides have been produced through enzymatic hydrolysis of soybean proteins. These peptides include ACE-inhibitory, antioxidative, opioid-like and bile acid binding peptides. Kuba *et al.* [83] reported that *Monascus purpureus*' acid protease hydrolysis of soybean proteins, β-conglycinin and glycinin, resulted in the production of four ACE inhibitory peptides with IC_{50} values of 65–850 μM. This research showed correlation between ACE inhibitory activity and the degree of hydrolysis. The active peptides have been identified as LAIPVNKP, LPHF, SPYP, and WP. Further study using synthesized analogues of WL and LAIPVNKP confirm high ACE inhibitory activities of these peptides [83] hence offer promising application of the hydrolysate.

In another study using Alcalase and Flavourzyme™, Chiang *et al.* [84] reported that a 6 h hydrolysis of soy protein with Alcalase would produce hydrolysate with highest ACE inhibitory activity when compared to hydrolysates of Flavourzyme™, trypsin, chymotrypsin, and pepsin. It was also evident that the higher the degree of hydrolysis (DH), the better will be the ACE inhibitory activity [84]. Hence, increase in the hydrolysis time indicated the significance of protease hydrolysis of proteins to produce bioactive peptides. In case of Flavourzyme™, however, the increase in degree of hydrolysis did not guarantee the increase in ACE inhibitor activity [84]. This might be due to inactivation of active peptide sequence by exoproteases present in Flavourzyme™, through the release of one or more amino acids from the active N- or C-terminal of the peptides. This is, however, in direct contrast to the finding of Suh *et al.* [85] which showed that Flavourzyme™ hydrolysates of corn gluten had higher ACE inhibitory activity. This suggests that different proteins may produce different bioactive peptides when hydrolysed with the same enzyme.

Many other researchers have reported the release of ACEIPs from hydrolysis of soybean proteins (Table 7.4). For example, recently ACEIPs have been identified from PROTIN SD-NY10 hydrolysed soy milk and soy protein containing infant formulas [27, 86]. The identification of ACEIPs from PROTIN SD-NY10 processed soy milk may have significant commercial implication as soy milk usually has low sensory acceptability among western consumers due to its beany flavor. Hence, hydrolysis may improve its acceptability, the produced ACEIPs and the associated health benefits may be available to the wider community. The finding of native ACEIP RPSYT in soybean-based infant formula [27] is fascinating. A closer look at the sequences of soy storage proteins reveals that this pentapeptide comes from the sequence of the glycinins (G1, G2, and G3) with homologue also exists in G5 chain. Therefore, it can be concluded that hydrolysed soybean proteins may form the most suitable ingredients in the formulation of infant foods.

Hydrolysis of soybean proteins has also resulted in the production of antioxidative peptides. For example, Moure *et al.* reported the antioxidative effect of ultrafiltration-recovered soy protein fractions from industrial effluents and their hydrolysates [87]. In their study, high molecular weight fractions (>10 kDa) of hydrolysates and proteins have been found to show higher antioxidant activities when compared to the lower molecular weight fractions [87], although the sequences of active peptides have not been determined. Other researchers, however, have reported various antioxidative peptides derived through hydrolysis of soy proteins [2, 53, 88]. Chen *et al.* reported six antioxidative peptides derived from *Bacillus sp.* Protease S hydrolysis of β-Conglycinin [53]. Three of these peptides, that is, LLPHH, VNPHDHQN, and LVNPHDHQN are potential antioxidative peptides that can extend the induction period of linoleic acid oxidation [53] and hence possess potential application in food industry.

In another study, pancreatic trypsin hydrolysis of β-Conglycinin and glycinin has resulted in the production of antioxidative peptides [2]. These peptides come from fragments of β-conglycinin and glycinin and their antioxidative activities have been characterized by Trolox equivalent antioxidative capacity (TEAC). Of these peptides, TTYY has been considered as the most promising peptide that can be used as functional compound in nutrition, pharmaceutical, and cosmetic applications [2]. The antioxidative peptides derived from soybean proteins reported so far come from the storage proteins. However, there is also antioxidative peptide that has been derived from other soy proteins. Jiménez-Escrig *et al.* reported the isolation of antioxidative peptide TIIPLPV derived from peptic hydrolysis of lipoxygenase-1 found in okara [88]. Okara is a by-product of soymilk and tofu industry and is a rich source of proteins. This finding serves a way to utilize sources that are otherwise wasted and convert them into functional ingredients.

Other functional properties of significance are immune modulating, hypocholesterolemic, and anticancer properties of soy peptides. Yoshikawa *et al.* reported the finding of a hexapeptide HCQRPR and a tetrapeptide QRPR derived from glycinin G1 through tryptic hydrolysis [3]. These peptides have shown to stimulate phagocytosis by microphages and polymorpho nuclear leucocytes [3]. Tryptic hydrolysis of soy protein β-conglycinin α' subunit has also resulted in the release of three active immune modulating peptides known as Soymetide-13 (MITLAIPVNKPGR), Soymetide-9 (MITLAIPVN), and Soymetide-4 (MITL) (Table 7.4) [89, 90]. These peptides have shown to stimulate phagocytosis by human polymorphonuclear leucocytes [89]. The soymetides have shown weak affinity toward the formyl peptide receptor, the receptor that detects bacterial infection, hence consumption of soy proteins may stimulate human immune system as though a bacterial infection has occurred [31]. Soymetide-4 has not shown to induce reactive oxygen species (ROS) property is differentto formyl-MLP (fMLP), a formylated immunomodulating peptide containing essential methionine (M) residue, and can induce the release of tumor necrosis factor α (TNF-α). Therefore, soymetides-4 is considered as a safe immunomodulating peptide [31].

The hydrolysis of soybean proteins, as discussed above, results in the release of significant numbers of bioactive peptides. These peptides possess various functional properties such as ACE-inhibitory, antioxidative and immunomodulating activities and possibly other functionalities. These peptides have shown to exert physiological activities *in vitro* and some of these have shown activity *in vivo* on animal models. Therefore, these peptides offer significant prospect for application in food and pharmaceutical industry.

7.4.3 Bioactive Peptides Derived Through Chemical Synthesis

Chemical synthesis of peptides of known sequences and for desired functionalities is an alternative for the production of large quantity of pure active peptides. There are two main methods for chemical synthesis of active peptides, that is, solution- and solid-phase syntheses. In the solution-phase synthesis, stepwise synthesis of small peptide fragments is carried out first followed by coupling the small peptides into longer peptides [96]. This method has been used successfully to synthesize more than 20 residue peptides. However, this method is time-consuming and the intermediate

Table 7.4 Bioactive peptides derived from soybean proteins by different digestion procedures.

Peptide sequences	Protein source	Activity*	Method of digestion	Reference
Angiotensin I-converting enzyme (ACE) inhibitors				
NWGPLV	Glycinin G4	21.0	Protease D3 hydrolysis	[31]
YVVFK	β-Conglycinin and glycinin	44.0	Protease D3 hydrolysis	[31]
PNNKPFQ	Glycinin G1, G2, and G3	33.0	Protease D3 hydrolysis	[31]
IPPGVPYWT	Glycinin G4	64.00	Protease D3 hydrolysis	[31]
TPRVF	Glycinin G4	200.0	Protease D3 hydrolysis	[31]
PGTAVFK	Glycinin G4	26.50	*B. subtilis* protease hydrolysis	[91]
FFL	β-Conglycinin α-chain	37.0	Peptic hydrolysis	[1]
IYLL	β-Conglycinin and glycinin	42.0	Peptic hydrolysis	[1]
IA	β-Conglycinin and glycinin	153.0	Peptic hydrolysis	[1]
YLAGNQ	Glycinin G1, G2, and G3	14.0	Peptic hydrolysis	[1]
VMDKPQG	–	39.0	Peptic hydrolysis	[1]
DLP	β-Conglycinin and glycinin	4.8	Alkaline hydrolysis	[92]
DG	Glycinin G1, G2, G3, and protease inhibitor	12.30	Alkaline hydrolysis	[92]
TIIPLPV	Glycinin G7	–	GIE hydrolysis	[88]
HHL	β-Conglycinin and glycinin	2.2	Fermentation	[93]
LIVTQ	Glycinin G1, G2, G3	0.09	LAB fermentation	[66]
LSW	Glycinin G1 and G3	2.70	Thermolysin hydrolysis	[94]

(Continued)

Table 7.4 Cont.

Peptide sequences	Protein source	Activity*	Method of digestion	Reference
IVF	Glycinin G4 and G7	63.30	Thermolysin hydrolysis	[94]
LNF	β-Conglycinin α- and β-chain	511.40	Thermolysin hydrolysis	[94]
LEF	Glycinin G1	655.20	Thermolysin hydrolysis	[94]
FFYY	–	1.90	PROTIN SD-NY10 hydrolysis	[86]
WHP	–	4.80	PROTIN SD-NY10 hydrolysis	[86]
FVP	Glycinin G1, G2, G3	10.10	PROTIN SD-NY10 hydrolysis	[86]
LHPGDAQR	β-Conglycinin	10.30	PROTIN SD-NY10 hydrolysis	[86]
IAV	Glycinin G1, G2, G3	27.00	PROTIN SD-NY10 hydrolysis	[86]
LEPP	Glycinin G4	100.10	PROTIN SD-NY10 hydrolysis	[86]
WNPR	Glycinin G1, G2, G3	880.0	PROTIN SD-NY10 hydrolysis	[86]
RPSYT	Glycinin G1, G2, G3	393.0	Native in Infant formula	[27]
LAIPVNKP	β-Conglycinin	70.0	Acid proteinase hydrolysis	[83]
LPHF	β-Conglycinin and glycinin	670.0	Acid proteinase hydrolysis	[83]
SPYP	β-Conglycinin and glycinin	850.0	Acid proteinase hydrolysis	[83]
WP	β-Conglycinin and glycinin	65.0	Acid proteinase hydrolysis	[83]
Antioxidative Peptides				
LLPHH	β-Conglycinin	–	Protease S hydrolysis	[53]
VNPHDHQN	β-Conglycinin	–	Protease S hydrolysis	[53]

LLPHHADADY	*β*-Conglycinin	–	Protease S hydrolysis	[53]
VIPAGYP	*β*-Conglycinin	–	Protease S hydrolysis	[53]
LVNPHDHQN	*β*-Conglycinin	–	Protease S hydrolysis	[53]
LNSGDALRVPSGTTYY	*β*-Conglycinin	–	Protease S hydrolysis	[53]
TTYY	*β*-Conglycinin	–	Trypsin hydrolysis	[2]
LY	*β*-Conglycinin and glycinin	–	Trypsin hydrolysis	[2]
IY	*β*-Conglycinin and glycinin	–	Trypsin hydrolysis	[2]
TIIPLPV	Lipoxygenase-1	–	GIE hydrolysis	[88]
Opioid Peptides				
YPFVV	*β*-Conglycinin	6.0	Chemical synthesis	[95]
YPFVVN	*β*-Conglycinin	9.20	Chemical synthesis	[95]
YPFVVNA	*β*-Conglycinin	13.0	Chemical synthesis	[95]
Immunomodulating Peptides				
MITLAIPVNKPGR	*β*-Conglycinin α′ subunit	50.0	Tryptic hydrolysis	[89]
MITLAIPVN	*β*-Conglycinin α′ subunit	25.0	Tryptic hydrolysis	[89]
MITL	*β*-Conglycinin α′ subunit	450.0	Tryptic hydrolysis	[89]
HCQRPR	Glycinin G1	–	Tryptic hydrolysis	[3]
QRPR	Glycinin G1	–	Tryptic hydrolysis	[3]
Bile-acid Binding Peptides				
VAWWMY	Glycinin G1 and G2	–		[4]

*The activityis expressed as concentration half maxima (IC_{50}) values (μM).

products need to be isolated and purified before proceeding to the next steps [96]. In the solid-phase synthesis, an α-amino group and side chain are protected and the protection groups are immobilized into a solid resin. The protected α-amino group can be removed quickly and completely, and another protected amino acid with activated carboxyl group can be coupled to the unprotected resin-bound amine [97]. This step can be repeated until a desired sequence is obtained.

In addition to the above two methods, active peptides can be synthesized though enzymatic process. This method is the opposite of enzymatic hydrolysis and is carried out in nonaqueous solutions. This method can be organized into thermodynamically and kinetically controlled processes [7]. At present, solid-phase peptide synthesis is considered as a more established method and has been used to synthesize significant numbers of active peptides. There are various reasons to produce bioactive peptides through chemical synthesis. These include producing peptides of high purity and quantity, peptides with desired sequences that are otherwise difficult to obtain from natural sources, peptides of known activity identified in natural sources but are difficult to ensure the release of predicted active peptides that are encrypted in native proteins through hydrolysis. Hence chemical synthesis provides alternative way to produce active peptides for further studies as well as for commercial application.

Various active peptides have been produced through chemical synthesis. Ohinata *et al.* had synthesized three peptides (YPFVV, YPFVVN, and YPFVVNA) identified in the sequence of *β*-Conglycinin (Table 7.4). The peptides have then been subjected for *in vitro* analysis for their opioid-like activity hence the reported finding of opioid peptides from the fragments of *β*-conglycinin [95]. This approach has also enabled the researchers to identify the most prospective peptide for further study and application.

The hypocholesterolemic effect of bioactive peptides derived from soy proteins has also been the subject research using synthetic peptides. It has been claimed that soy proteins offer highest hypocholesterolemic property when compared to other proteins [4, 8, 90]. For example, using information of the bile acid-binding region of soy glycinin G1 subunit identified by Choi *et al.* [8], Nagaoka *et al.* [4] has synthesized a hexapeptide VAWWMY and studied its bile acid-binding activity *in vivo*. The results from this study confirm that VAWWMY (Table 7.4) does exert bile acid-binding ability by decreasing micellar solubility and inhibiting cholesterol absorption in rats [4] hence showing hypocholesterolemic property.

The above are only a few examples of the application of chemical synthesis of active peptides. This approach can be used to produce bioactive peptides of all physiological functionalities such as antioxidative peptides [98], ACE-inhibitory peptides [34, 32] and antimicrobial peptides [99, 100].

7.4.4 Perspectives in Bioactive Peptide Production Through Recombinant DNA Technology

The three methods discussed above are the most commonly used in preparation of bioactive peptides. The isolation of the peptides from the fermented products or the crude hydrolysates is time consuming and expensive as the products contain a mixture of numerous peptides of different properties and of low quantity. On the other hand, chemical synthesis is an expensive process. Recombinant DNA technology has been used to produce bioactive peptides derived from food proteins. For example, ACEIPs derived from bovine β-casein [9] and human α_{S1}-casein [101]. This method aims to produce proteins containing selected bioactive peptides of known sequences. The active peptides can then be released through appropriate enzymatic hydrolysis.

The recombinant DNA technology involves the use of a cloning vector that is derived from organisms such as bacteria or viruses. The most commonly used cloning vector is *Escherichia coli*. For example, Prak *et al.* introduced tandem multimers of the nucleotide sequence encoding IIAEK peptide (an active hypocholesterolemic peptide derived from β-lactoglobulin) into DNA regions corresponding to the five variable regions of soybean glycinin A1aB1b (glycinin G1) subunit and expressed the mutants in *E. coli* [102]. This introduction is followed by screening of the expressed mutant proteins containing the bioactive peptides that are most soluble in *E. coli* and used as a selected mutant for further protein production. The proteins are then isolated and hydrolysed with trypsin to release the active peptides [102]. This approach has resulted in the production of multi copies of IIAEK peptide from this genetically modified soybean pro-glycinin A1aB1b (pro-glycinin G1).

The use of *E. coli* strains as cloning vectors have been common in the literature [101, 103]. However, these strains are not food grade strains hence the need to isolate the recombinant proteins for further use. More recently, organisms that are Generally Recognized as Safe (GRAS) by the Food and Drug Administration of the United States of America have been used for this purpose. These organisms include lactic acid bacteria (LAB) *Bifidobacterium pseudocatenulatum* [104] and *Chlamydomonas reinhardtii* [105]. Losurdo *et al.* (2013) used *E. coli* and *B. pseudocatenulatum* as shuttle vector [104] to produce recombinant protein containing ACEIPs previously identified in the sequence of bovine β-casein [106].

In this study, positive clone carrying the recombinant clones is introduced into pAM1 plasmid form *E. coli-bifidobacteria* shuttle cloning vector [107, 108] then introduced into *E. coli* for further transformation. Positive recombinant plasmids carrying in-frame genes are then electrotransformed into *B. pseudocatenulatum* for further growth. The bacteria are then harvested and the proteins are extracted and used for ACE inhibitory analysis [104].

Another species of interest as cloning vector is *Chlamydomonas reinhardtii*, a single-cell green algae. This organism has been used to produce therapeutic proteins in its chloroplast and is considered an inexpensive and easily scalable cloning system for quantity recombinant protein production [105]. To date, genetic engineering in chloroplast of *C. reinhardtii* has mainly been limited to the production of proteins with therapeutic applications [105]. An extended list of therapeutic proteins produced through genetic engineering using *C. reinhardtii* is available in the literature [105]. The principle of therapeutic protein production in *C. reinhardtii* chloroplast involves the introduction of gold or tungsten particles coated with gene for the foreign protein into the algae cells placed at an interior of a vacuum chamber [105, 109]. This step is carried out by a particle bombardment device. The cells are then allowed to proliferate for a few weeks in the presence of selected antibiotic. When the genes reached the chloroplast, they will integrate into the plastid (plant plasmid) and in the chloroplast genome [105, 109].

As mentioned earlier, genetic engineering using *C. reinhardtii* has been mainly for the production of therapeutic proteins. Other potential cloning vectors for genetic engineering are transgenic flowering plants. These plants have been used mainly for the production of oils, pigments, enzymes, and vitamins, or for the innovation of drought and salt tolerant plants [105, 110, 111]. Plant cells contain DNA in their nucleus, plastid and mitochondrial compartments, of which the nuclear genome is most common target for genetic engineering. However, the plastids (in this case the chloroplast) in plants have also been the target for successful expression of antibiotic protein [110]. In light of this successful report, genetic manipulation though recombinant DNA technology for the purpose of bioactive peptide production is feasible. The production of recombinant proteins carrying bioactive peptides through transgenic plants offer several benefits: (1) scalable production can be achieved once a positive clone and farming conditions have been established, (2) the recombinant proteins can be used in food preparation, (3) the recombinant proteins can be extracted, hydrolyzed with appropriate enzyme to release the active peptides that can be used as therapeutic agents, and (4) in case recombinant antibiotic proteins in which mechanism of action involves microbial membrane disruption, in transgenic plant production offer additional benefits [113] as chloroplasts do not have cell walls hence can accommodate such proteins and chloroplast have similar proteases as those of bacteria hence will express high resistance toward their degradation.

The above discussions do indicate that recombinant DNA technology offers beneficial alternative for the production of bioactive peptides and proteins:

- Recombinant proteins containing encrypted ACEIPs and cholesterol lowering peptides have been successfully produced through *E. coli* as the cloning vector and *E. coli-bifidobacteria* as shuttle vector.
- Generally Recognized as Safe (GRAS) cloning vectors are now available for application. These include LAB *B. pseudocatenulatum*, single-cell *C. reinhardtii* and transgenic flowering plants. The latter two offer advantages for scalable production of therapeutic proteins and in particular for membrane disrupting antibiotic proteins and peptides.
- The recombinant active proteins produced by GRAS cloning vector can be used for food preparation.
- At the present time, recombinant DNA technology for protein production remains in its infancy. However, current research findings do indicate that the technology for the production of recombinant proteins containing encrypted bioactive peptides is feasible and hence further research is required to fully investigate the best cloning vehicles to produce recombinant proteins containing many different bioactive peptide classes.

7.4.5 Drawbacks in Bioactive Peptide Production

Review presented in the previous sections does indicate the promising physiological benefits of peptides derived from soy protein degradations either through fermentation or enzymatic hydrolysis. However, production of peptides of specific physiological properties and their efficacy rely on many factors that may not be easily achieved. These factors include amino acid sequences of the parent proteins, appropriate enzymes to cleave the proteins, quantity of the active peptides produced, isolation of the active peptides, efficacy of the peptides, possible allergic reactions, and production cost.

The amino acid sequences of parent proteins are a determining factor in the release of bioactive peptides as these sequences may have variations between same proteins from different species or even between variant of a same species. The limited knowledge of the amino acid sequences of many food proteins means that the production of bioactive peptides from food proteins is largely a functional-guided trial-and-error experimentation using various food proteins and enzymes. The well-studied milk proteins, soy protein, blue lupin proteins and amaranth proteins are exception, where the well documented sequence can be used to guide the choice of enzymes needed to cleave the peptide bonds and release desired bioactive peptides. For these proteins, an *in silico* digestion can shed light on the appropriate enzymes to be used for actual hydrolysis of the proteins. The knowledge of enzyme specificity, therefore, plays an important role in achieving optimum condition for hydrolysis.

The quantity of active peptides produced from hydrolysis of food proteins depends on the number of occurrences of the desired peptide sequence within the parent proteins and the specificity of the enzyme used in hydrolysis. The yield of a particular active peptide is usually low. In addition to the low yield, purification of active peptides is a major challenge and is costly. At laboratory level, chromatographic separation is the most common method of choice. However, this method is costly and time consuming, and since the hydrolysates from hydrolysis contain numerous peptides with relatively similar in sizes and hydrophobicity, and contaminations are unavoidable. Various studies indicated that most bioactive peptides, in particular antihypertensive peptides, are small peptides and hence fractionation of hydrolysates with molecular weight cut off membrane to produce hydrolysate fraction containing active peptides is commendable and the fraction can be used in food product development. Production of recombinant proteins containing repeated sequences of active peptides and chemical synthesis of active peptides can improve the quantity of active peptides produced. The former method offers superior alternative as the peptides produced can be used as both nutraceuticals and therapeutic agents, while the latter is resorted to therapeutic application.

The efficacy of the active peptides and their potential cause of allergic reactions are other issues associated with the use of bioactive peptides. Reports from various studies indicated that there exists bioactive peptide that showed *in vitro* activity but failed when tested *in vivo*. These are believed to be the inability of these peptides to resist degradation by gastrointestinal proteases and other pancreatic enzymes. Other reports, however, showed that bioactive peptides can be absorbed intact into the blood circulation following an oral administration. Coming to the question of possible allergic reaction due to consumption of bioactive peptides, although, as far as recent information is concerned, there is no report of allergic reaction experienced during oral administration of bioactive peptides, antihypertensive peptides in particular, among volunteer subjects involved in bioactive peptide research.

7.5 Potential Applications of Bioactive Peptides from Soy Proteins

7.5.1 Applications in Functional Foods

Studies on bioactive peptides derived from food proteins indicate that there are many possible uses of these peptides. Two important ones are antihypertensive and antimicrobial agents. As an antihypertensive agent, the use of bioactive peptides as a food ingredient is advisable [112]. Therefore, the functional properties of these peptides assume importance.

According to Kinsella [113], properties such as aroma and taste are sometimes also regarded as functional. This is particularly important when these properties are exploited in the formulation of foods. In the case of enzymatic-modified food protein, Adler-Nissen [36] suggested that it is the off-flavor which dominates and hinders the use. Hence proposed an application in small quantity as highly functional ingredients in complex food items.

Various proteolytic enzymes have been used successfully to hydrolyse food proteins yielding different bioactive peptides with potential use as food ingredients. However certain proteases of bacterial origin (e.g., alkaline protease) need further investigation into antigen-antibody reaction (food allergy) and to digestive resistance [112]. This therefore indicates the need for *in vivo* investigations with laboratory animals and human studies to examine possible allergic reaction upon administration of these peptides.

For example, numerous *in vitro* studies have confirmed the activities of ACEIPs. *In vivo* studies with both animal and human subjects have also been carried out and promising results have been reported. Peptides IPP and VPP derived by *Lactobacillus helveticus* fermentation of milk, such as Calpis sour milk, are well characterized for their activities against ACE both *in vitro* and *in vivo* with animal and human subjects. A variety of products are available commercially either as functional food supplements or as bacterial-based functional foods. Among these are three milk-based functional foods with claimed potency for reducing blood pressure. They are Ameal S from Calpis (Kanagawa, Japan), Evolus from Valio (Finland), and Cardi-04™ from Chr. Hansen (Denmark). The first two products contain bioactive peptides IPP and VPP and some other active peptides. Cardi-04™ is a fermented milk product produced by *Lactobacillus helveticus* strain Cardi-04TM [114]. Fermented soybean products such as natto, miso and tempeh are staple foods in Japan and Indonesia. These food items have been demonstrated to possess ACE-inhibitory, antioxidative and anti-thrombotic properties [68, 115] and hence are functional food ingredients on their own rights.

Natto, miso, and tempeh are products from solid-state fermentation and contain low moisture content hence can be easily converted into dry food ingredients and be used in functional food preparation. Incorporation of these ingredients in food product development is feasible. For example in the preparation of functional breakfast cereal containing ACE inhibitory, cholesterol lowering and antioxidative peptides; soymilk enriched with isolated bioactive peptides and in the preparation of many other food items. In case of isolated antioxidative peptides such as LLPHH, VNPHDHQN, and LVNPHDHQN [53], their application in cooking oils to extend the induction period will greatly increase their shelf life. The application of these active peptides in food preparation is limitless and further product development experiments are required to maximize the use of these active ingredients.

7.5.2 Application as Therapeutic Agents

The use of bioactive peptides as therapeutic agents is one of the main objectives of developing these active constituents. For example, an antimicrobial agent derived from soybean fermentation may be used for clinical purposes and also as food preservative. According to Wilcox [116] to become a good therapeutic agent, a drug has to show good activity, appropriate function, low toxicity, and have stability *in vivo*. So far only two bioactive (cationic) peptides are used in topical creams and solutions. As a food preservative, Chan and Li-Chan [117] suggested that apart from having good *in vivo* activity, synergistic effect of an antimicrobial peptide with other preservative agents may be important for their successful use in food products.

Several ACEIPs derived from soy proteins (Table 7.4) are potential candidates for antihypertensive agents due to their small sizes and low IC_{50} values. These include DLP (IC_{50} = 4.8 μM), HHL (IC_{50} = 2.2 μM), LIVTQ (IC_{50} = 0.9 μM), LSW (IC_{50} = 1.70 μM), FFYY (IC_{50} = 1.90 μM), WHP (IC_{50} = 4.80 μM), and FVP (IC_{50} = 10.10 μM). As discussed in Section 7.3, these ACEIPs possess most of the structural properties required to bind with ACE active site and hence observed antihypertensive activities are expected.

Other soy protein derived small bioactive peptides, that is, antioxidative and immunomodulating peptides (Table 7.4), also have potential to be used as therapeutic agents. *In silico* hydrolysis of these peptides revealed that the peptides LLPHH, TIPLPV, HCQRPR, and QRPR are stable against gastrointestinal enzymes pepsin, chymotrypsin and trypsin degradation and hence are expected to deliver activity *in vivo*. Therefore, these peptides may be used in the preparation of health supplements or as active ingredients in topical cream products. The application of antioxidative peptides in topical cream may have twofold benefits, that is, to protect other oxidation prone ingredients and to deliver antioxidants to human skin [2]. Likewise, immunomodulating peptides can be incorporated into topical creams to treat skin inflammation and this has been demonstrated in the literature [118].

7.6 Conclusion

Soybean is a well-known food item that is consumed all over the world. The nutritive value of soybean protein is well recognized and is comparable to other food proteins. In addition to nutritional value, the soybean proteins exhibit physiological activities directly and also indirectly by release of active peptides from the gastrointestinal enzyme degradation as well as through isolated enzyme degradation *in vitro*and fermentation. As can be seen from the literature reviewed in this chapter, fermentation and enzymatic hydrolysis of soybean proteins result in the release of peptides with substantial physiological activities.

This chapter has reviewed several approaches available in the literature for producing bioactive peptides from soybean proteins that includes (1) fermentation using microbial enzymes, (2) *in vitro* hydrolysis by gastrointestinal and commercial enzymes, (3) chemical synthesis, and (4) recombinant DNA technology. The main emphasis of this review has been to identify various bioactive peptides derived from soybean proteins and to unravel their potential contribution to human health. It has been highlighted that an *in silico* bioinformatics approach can be employed to develop *in vitro* methodology to produce bioactive peptides through suitable enzyme hydrolysis. The use of protein and peptide online database such as BIOPEP in conjunction with QSAR models can then be used to identify and predict the existence of bioactive peptides in the simulated *in silico* hydrolysates. These approaches provide an insight into suitable enzymes needed to produce highly active peptides in maximum yield that can then be followed by enzymatic hydrolyses for the production and discovery of bioactive peptides. Peptides with antihypertensive, antioxidative, immunomodulatory, bile acid binding, opioid, and DPP-IV inhibitor activities have been identified by *in silico* method and predicted to have potential applications as nutraceuticals and therapeutic agents.

References

1. Chen, J.R., Okada, T., Muramoto, K., Suetsuna, K., Yang, S.C., Identification of angiotensin I-converting enzyme inhibitory peptides derived from the peptic digest of soybean protein. *J. Food Biochem.*, 26, 543–554, 2002.
2. Beermann, C., Euler, M., Herzberg, J., Stahl, B., Anti-oxidative capacity of enzymatically released peptides from soybean protein isolate. *Eur. Food Res. Technol.*, 229, 637–644, 2009.
3. Yoshikawa, M., Kishi, K., Takahashi, M., Watanabe, A., Miyamura, T., Yamazaki, M., Chiba, H., Immunostimulating peptide derived from soybean protein. *Ann. N. Y. Acad. Sci.*, 685, 375–376, 1993.
4. Nagaoka, S., Nakamura, Y.A., Shibata, H., Kanamaru, Y., Soystatin (VAWWMY), a novel bile acid-binding peptide, decreased micellar solubility and inhibited cholesterol absorption in rats. *Biosci., Biotechnol. Biochem.*, 74, 1738–1741, 2010.
5. Sanjukta, S., Rai, A.K., Production of bioactive peptides during soybean fermentation and their potential health benefits. *Trends Food Sci. Technol.*, 50, 1–10, 2016.
6. Salampessy, J., Reddy, N., 2016. *Unpublished Work.*
7. Guzmán, F., Barberis, S., Illanes, A., Peptide synthesis: chemical or enzymatic. *Electron. J. Biotechnol.*, 10, 279–314, 2007.
8. Choi, S.K., Adachi, M., Utsumi, S., Identification of the bile acid-binding region in the soy glycinin A1aB1b subunit. *Biosci., Biotechnol. Biochem.*, 66, 2395–2401, 2002.
9. Losacco, M., Gallerani, R., Gobbetti, M., Minervini, F., De Leo F., Production of active angiotensin-I converting enzyme inhibitory peptides derived from bovine beta-casein by recombinant DNA technologies. *Biotechnol J.* 2(11),1425–34, 2007.
10. Liu, K., *Soybeans as functional foods and ingredients*, AOCS press Champaign, IL, 2004.
11. Hettiarachchy, N.S., Kalapathy, U., Wu, M., Soybean protein: A Renewable Resource for Food and Non Food Ingredients. In: *Paradigm for successful utilization of renewable resources*, AOCS Publishing, 1998, pp. 183–197.
12. Egbert, W.R., Isolated Soy Protein: Technology, Properties, and Applications. In: *Soybeans as functional foods and ingredients*, AOCS Publishing, pp, 134–162, 2004.

13. Xiao, C.W., Functional soy products, In: *Functional foods*, 2nd edn., M. Saarela, Ed., Woodhead Publishing, 2011, pp. 534–556.
14. Young, V.R., Soy protein in relation to human protein and amino acid nutrition. *J. Am. Diet Assoc.*, 91, 828–835, 1991.
15. Fields, C., Mallee, P., Muzard, J., Lee, G.U., Isolation of Bowman-Birk-Inhibitor from soybean extracts using novel peptide probes and high gradient magnetic separation. *Food Chem.*, 134, 1831–1838, 2012.
16. Lin, P., Ng, T.B., A stable trypsin inhibitor from Chinese dull black soybeans with potentially exploitable activities. *Process Biochem.*, 43, 992–998, 2008.
17. McLaughlin, C.L., Peikin, S.R., Baile, C.A., Trypsin inhibitor effects on food intake and weight gain in Zucker rats. *Physiol. Behav.*, 31, 487–491, 1983.
18. McManus, M.T., Corner, R., Garthwaite, I., Characterization of monoclonal antibodies that recognize the soybean (Kunitz) trypsin inhibitor: binding to the inhibitor interrupts the formation of the trypsin: inhibitor complex. *J. Plant Physiol.*, 146, 243–248, 1995.
19. Sagili, R.R., Pankiw, T., Zhu-Salzman, K., Effects of soybean trypsin inhibitor on hypopharyngeal gland protein content, total midgut protease activity and survival of the honey bee (Apis mellifera L.). *J. Insect Physiol.*, 51, 953–957, 2005.
20. Bakke, A.M., Chikwati, E.M., Venold, F.F., Sahlmann, C., Holm, H., Penn, M.H., Oropeza-Moe, M., Krogdahl, Å., Bile enhances glucose uptake, reduces permeability, and modulates effects of lectins, trypsin inhibitors and saponins on intestinal tissue. *Comp. Biochem. Physiol. Part A: Mol. Integr. Physiol.*, 168, 96–109, 2014.
21. Chen, Y., Xu, Z., Zhang, C., Kong, X., Hua, Y., Heat-induced inactivation mechanisms of Kunitz trypsin inhibitor and Bowman-Birk inhibitor in soymilk processing. *Food Chem.*, 154, 108–116, 2014.
22. Johnson, L.A., Myers, D.J., Burden, D.J., Early uses of soy protein in far east. *Inform*, 3, 282–290, 1992.
23. Singh, B.P., Vij, S., Hati, S., Functional significance of bioactive peptides derived from soybean. *Peptides*, 54, 171–179, 2014.
24. Wang, Y.C., Yu, R.C., Chou, C.C., Antioxidative activities of soymilk fermented with lactic acid bacteria and bifidobacteria. *Food Microbiol.*, 23, 128–135, 2006.
25. Zhao, J., Xiong, Y.L., Interfacial peptide partitioning and undiminished antioxidative and emulsifying activity of oxidatively stressed soy protein hydrolysate in an O/W emulsion. *LWT - Food Sci. Technol.*, 61, 322–329, 2015.
26. De Mejia, E., De Lumen, B.O., Soybean bioactive peptides: A new horizon in preventing chronic diseases. *Sex. Reprod. Menopause*, 4, 91–95, 2006.
27. Puchalska, P., Concepción García, M., Luisa Marina, M., Identification of native angiotensin-I converting enzyme inhibitory peptides in commercial soybean based infant formulas using HPLC-Q-ToF-MS. *Food Chem.*, 157, 62–69, 2014.

28. Tsai, J.S., Lin, Y.S., Pan, B.S., Chen, T.J., Antihypertensive peptides and γ-aminobutyric acid from prozyme 6 facilitated lactic acid bacteria fermentation of soymilk. *Process Biochem.*, 41, 1282–1288, 2006.
29. Kim, S.K., Kim, H.H., Kim, J.Y., Kang, Y.I., Woo, H.J., Lee, S.E., Anticancer activity of hydrophobic peptides from soy proteins. *Biofactors*, 12, 151–155, 2000.
30. Goldflus, F., Ceccantini, M., Santos, W., Amino acid content of soybean samples collected in different Brazilian states: harvest 2003/2004. *Revista Brasileira de Ciência Avícola*, 8, 2006.
31. Kodera, T., Nio, N., Identification of an Angiotensin I-converting Enzyme Inhibitory Peptides from Protein Hydrolysates by a Soybean Protease and the Antihypertensive Effects of Hydrolysates in 4 Spontaneously Hypertensive Model Rats. , *Food Science* 71, C164–C173, 2006.
32. Li, F., Ohnishi-Kameyama, M., Takahashi, Y., Yamaki, K., Angiotensin I-converting enzyme inhibitory activities of Chinese fermented soypaste and estimation of the inhibitory substances. *J. Funct. Foods*, 5, 1991–1995, 2013.
33. Shin, Z.I., Yu, R., Park, S.A., Chung, D.K., Ahn, C.W., Nam, H.S., Kim, K.S., Lee, H.J., His-His-Leu, an angiotensin I converting enzyme inhibitory peptide derived from Korean soybean paste, exerts antihypertensive activity in vivo. *J. Agric. Food Chem.*, 49, 3004–3009, 2001.
34. Salampessy, J., Reddy, N., Kailasapathy, K., Phillips, M., Functional and potential therapeutic ACE-inhibitory peptides derived from bromelain hydrolysis of trevally proteins. *J. Funct. Foods,* 14, 716–725, 2015.
35. Dziuba, J., Iwaniak, A., Database of protein and bioactive peptide sequences. In: *Nutraceutical proteins and peptides in health and disease*, Y. Mine, F. Shahidi, Eds., CRC-Taylor and Francis, London, pp. 543–563, 2006.
36. Adler-Nissen, J., *Enzymic hydrolysis of food proteins*, Elsevier Applied Science Publishers, London, 1986.
37. Minkiewicz, P., Dziuba, J., Iwaniak, A., Dziuba, M., Darewicz, M., BIOPEP database and other programs for processing bioactive peptide sequences. *J. AOAC Int.*, 91, 965–980, 2008.
38. Banèeres, J.-L., Essalouh, L., Jariel-Encontre, I., Mesnier, D., Garrod, S., Parello, J., Evidence indicating proximity in the nucleosome between the histone H4 N termini and the globular domain of histone H1. *J. Mol. Biol.*, 243, 48–59, 1994.
39. Mitchell, W.M., Cleavage at arginine residues by clostripain. In: *Methods in enzymology*, Academic Press, Vol. 47, pp. 165–170, 1977.
40. Tarentino, A.L., Quinones, G., Grimwood, B.G., Hauer, C.R., Plummer, T.H., Molecular cloning and sequence analysis of flavastacin: an O-glycosylated prokaryotic zinc metalloendopeptidase. *Arch. Biochem. Biophys.*, 319, 281–285, 1995.
41. Keil, B., *Specificity of proteolysis*, Springer-Verlag, 1992.
42. Maruyama, N., Prak, K., Motoyama, S., Choi, S.K., Yagasaki, K., Ishimoto, M., Utsumi, S., Structure–Physicochemical Function Relationships of Soybean

Glycinin at Subunit Levels Assessed by Using Mutant Lines. *J. Agric. Food Chem.*, 52, 8197–8201, 2004.

43. Kamran, F., Salampessy, J., Reddy, N., *Application of NMR Spectroscopy for Structural Characterization of Bioactive Peptides Derived from food Protein.* In: *Application of NMR spectroscopy in food science.* F.R.S. Atta-ur-Rahman, M.I. Chaudhary, Eds., Bentham Science Publishers, 5, 3–76, London, 2016.
44. Beilinson, V., Chen, Z., Shoemaker, R., Fischer, R., Goldberg, R., Nielsen, N., Genomic organization of glycinin genes in soybean. *Theor. Appl. Genet.*, 104, 1132–1140, 2002.
45. Hernández-Ledesma, B., Hsieh, C.C., de Lumen, B.O., Lunasin and Bowman-Birk protease inhibitor (BBI) in US commercial soy foods. *Food Chem.*, 115, 574–580, 2009.
46. Ribeiro, J.K., Cunha, D.D., Fook, J.M., Sales, M.P., New properties of the soybean trypsin inhibitor: inhibition of human neutrophil elastase and its effect on acute pulmonary injury. *Eur. J. Pharm.*, 644, 238–244, 2010.
47. Nakamura, K., Naramoto, K., Koyama, M., Blood-pressure-lowering effect of fermented buckwheat sprouts in spontaneously hypertensive rats. *J. Funct. Foods*, 5, 406–415, 2013.
48. Ngo, D.H., Kang, K.H., Jung, W.K., Byun, H.G., Kim, S.K., Protective effects of peptides from skate (Okamejei kenojei) skin gelatin against endothelial dysfunction. *J. Funct. Foods*, 10, 243–251, 2014.
49. Balti, R., Bougatef, A., Sila, A., Guillochon, D., Dhulster, P., Nedjar-Arroume, N., Nine novel angiotensin I-converting enzyme (ACE) inhibitory peptides from cuttlefish (Sepia officinalis) muscle protein hydrolysates and antihypertensive effect of the potent active peptide in spontaneously hypertensive rats. *Food Chem.*, 170, 519–525, 2015.
50. Jao, C.L., Huang, S.L., Hsu, K.C., Angiotensin I-converting enzyme inhibitory peptides: inhibition mode, bioavailability, and antihypertensive effects. *BioMedicine*, 2, 130–136, 2012.
51. Wu, J., Aluko, R.E., Nakai, S., Structural requirements of angiotensin I-converting enzyme inhibitory peptides: quantitative structure–activity relationship study of di-and tripeptides. *J. Agric. Food Chem.*, 54, 732–738, 2006.
52. Foltz, M., Meynen, E., Bianco, V., van Platerink, C., Angiotensin converting enzyme inhibitory peptides from a lactotripeptide-enriched milk beverage are absorbed intact into the circulation. *J. Nutr.*, 137, 953, 2007.
53. Chen, H.M., Muramoto, K., Yamauchi, F., Structural analysis of antioxidative peptides from Soybean. beta.-Conglycinin. *J. Agric. Food Chem.*, 43, 574–578, 1995.
54. Saito, K., Jin, D.H., Ogawa, T., Muramoto, K., Hatakeyama, E., Yasuhara, T., Nokihara, K., Antioxidative properties of tripeptide libraries prepared by the combinatorial chemistry. *J. Agric. Food Chem.*, 51, 3668–3674, 2003.
55. Jao, C.L., Hung, C.C., Tung, Y.S., Lin, P.Y., Chen, M.C., Hsu, K.C., The development of bioactive peptides from dietary proteins as a dipeptidyl peptidase

IV inhibitor for the management of type 2 diabetes. *BioMedicine*, 5, 1–7, 2015.

56. Lacroix, I.M.E., Li-Chan, E.C.Y., Food-derived dipeptidyl-peptidase IV inhibitors as a potential approach for glycemic regulation–Current knowledge and future research considerations. *Trends Food Sci. Technol.*, 54, 1–16, 2016.
57. Patil, P., Mandal, S., Tomar, S.K., Anand, S., Food protein-derived bioactive peptides in management of type 2 diabetes. *Eur. J. Nutr.*, 54, 863–880, 2015.
58. Nongonierma, A. B., FitzGerald, R. J., Prospects for the management of type 2 diabetes using food protein-derived peptides with dipeptidyl peptidase IV (DPP-IV) inhibitory activity. *Curr. Opin. Food Sci.*, 8, 19–24, 2016.
59. Natesh, R., Schwager, S.L.U., Sturrock, E.D., Acharya, K.R., Crystal structure of the human angiotensin-converting enzyme–lisinopril complex. *Nature*, 421, 551–554, 2003.
60. Ariyoshi, Y., Angiotensin-converting enzyme inhibitors derived from food proteins. *Treands Food Sci. Technol.*, 4, 139–144, 1993.
61. Cheung, H.S., Wang, F.L., Ondetti, M.A., Sabo, E.F., Cushman, D.W., Binding of peptide substrates and inhibitors of angiotensin-converting enzyme. Importance of the COOH-terminal dipeptide sequence. *J. Biol. Chem.*, 255, 401–407, 1980.
62. Peredes-Lopez, O., Harry, G.I., Food biotechnology review: Traditional solid-state fermentations of plant raw materials—application, nutritional significance, and future prospects. *Crit. Rev. Food Sci. Nutr.*, 27, 159–187, 1988.
63. Hesseltine, C.W., Some important fermented foods of Mid-Asia, the Middle East, and Africa. *J. Am. Oil Chem. Soc.*, 56, 367–374, 1979.
64. Winarno, F.G., Fermented vegetable protein and related foods of Southeast Asia with special reference to Indonesia. *J. Am. Oil Chem. Soc.*, 56, 363–366, 1979.
65. Pihlanto, A., Korhonen, H., 3–Bioactive peptides from fermented foods and health promotion A2. In: *Advances in fermented foods and beverages*, W. Holzapfel, Ed., Woodhead Publishing, pp. 39–74, 2015.
66. Vallabha, S. V., Tiku, P.K., Antihypertensive peptides derived from soy protein by fermentation. *Int. J. Pept. Res. Ther.*, 20, 161–168, 2014.
67. Rho, S.J., Lee, J.S., Chung, Y.I., Kim, Y.W., Lee, H.G., Purification and identification of an angiotensin I-converting enzyme inhibitory peptide from fermented soybean extract. *Process Biochem.*, 44, 490–493, 2009.
68. Gibbs, B.F., Zougman, A., Masse, R., Mulligan, C., Production and characterization of bioactive peptides from soy hydrolysate and soy-fermented food. *Food Res. Int.*, 37, 123–131, 2004.
69. Amadou, I., Gbadamosi, O.S., Shi, Y.H., Kamara, M.T., Jin, S., Le, G.W., *Res. J. Microbiol.*, 5, 372–380, 2010.
70. de Souza Ferreira, E., Silva, M.A., Demonte, A., Neves, V.A., β-Conglycinin (7S) and glycinin (11S) exert a hypocholesterolemic effect comparable to that

of fenofibrate in rats fed a high-cholesterol diet. *J. Funct. Foods*, 2, 275–283, 2010.

71. Zhong, F., Liu, J., Ma, J., Shoemaker, C.F., Preparation of hypocholesterol peptides from soy protein and their hypocholesterolemic effect in mice. *Food Res. Int.*, 40, 661–667, 2007.
72. Maki, K.C., Butteiger, D.N., Rains, T.M., Lawless, A., Reeves, M. S., Schasteen, C., Krul, E.S., Effects of soy protein on lipoprotein lipids and fecal bile acid excretion in men and women with moderate hypercholesterolemia. *J. Clin. Lipidol.*, 4, 531–542, 2010.
73. Wang, D., Wang, L.J., Zhu, F.X., Zhu, J.Y., Chen, X.D., Zou, L., Saito, M., Li, L.T., In vitro and in vivo studies on the antioxidant activities of the aqueous extracts of Douchi (a traditional Chinese salt-fermented soybean food). *Food Chem.*, 107, 1421–1428, 2008.
74. Kobayashi, M., Hirahata, R., Egusa, S., Fukuda, M., Hypocholesterolemic effects of lactic acid-fermented soymilk on rats fed a high cholesterol diet. *Nutrients*, 4, 1304–1316, 2012.
75. Sites, C.K., Cooper, B.C., Toth, M.J., Gastaldelli, A., Arabshahi, A., Barnes, S., Effect of a daily supplement of soy protein on body composition and insulin secretion in postmenopausal women. *Fertil. Steril.*, 88, 1609, 2007.
76. Kwon, D.Y., Hong, S.M., Ahn, I.S., Kim, M.J., Yang, H.J., Park, S., Isoflavonoids and peptides from meju, long-term fermented soybeans, increase insulin sensitivity and exert insulinotropic effects in vitro. *Nutrition*, 27, 244–252, 2011.
77. Lee, M.H., Lee, J., Nam, Y.D., Lee, J.S., Seo, M.J., Yi, S.H., Characterization of antimicrobial lipopeptides produced by Bacillus sp. LM7 isolated from chungkookjang, a Korean traditional fermented soybean food. *Int. J. Food Microbiol.*, 221, 12–18, 2016.
78. Yeo, I.C., Lee, N.K., Cha, C.J., Hahm, Y.T., Narrow antagonistic activity of antimicrobial peptide from Bacillus subtilis SCK-2 against Bacillus cereus. *J. Biosci. Bioeng.*, 112, 338–344, 2011.
79. Shon, D.H., Lee, K.E., Ahn, C.W., Nam, H.S., Lee, H.J., Shin, J.K., Screening of antithrombotic peptides from soybean paste by the microplate method. *Korean J. Food Sci. Technol.*, 28, 684–688, 1996.
80. Whitaker, J.R., Protein Structure and Kinetics of Enzyme Reactions. In: *Handbook of food enzymology*, Whitaker, J.R., Voragen, A.G.J., Wong, D.W.S., Marcel Dekker Inc., New York, 2003a.
81. Whitaker, J.R., Proteolytic Enzymes. In: *Handbook of food enzymology*, J.R. Whitaker, A.G.J. Voragen, D.W.S. Wong, Eds., Marcel Dekker Inc., New York, 2003b.
82. Barret, A.J., Proteolytic enzymes: nomenclature and classification. In: *Proteolytic enzymes*, R. Beynon, J.S. Bond, Eds., Oxford University Press, New York, 2nd edn., 2001.
83. Kuba, M., Tana, C., Tawata, S., Yasuda, M., Production of angiotensin I-converting enzyme inhibitory peptides from soybean protein with Monascus purpureus acid proteinase. *Process Biochem.*, 40, 2191–2196, 2005.

84. Chiang, W.D., Tsou, M.J., Tsai, Z.Y., Tsai, T.C., Angiotensin I-converting enzyme inhibitor derived from soy protein hydrolysate and produced by using membrane reactor. *Food Chem.*, 98, 725–732, 2006.
85. Suh, H.J., Whang, J.H., Kim, Y.S., Bae, S.H., Noh, D.O., Preparation of angiotensin I converting enzyme inhibitor from corn gluten. *Process Biochem.*, 38, 1239–1244, 2003.
86. Tomatsu, M., Shimakage, A., Shinbo, M., Yamada, S., Takahashi, S., Novel angiotensin I-converting enzyme inhibitory peptides derived from soya milk. *Food Chem.*, 136, 612–616, 2013.
87. Moure, A., Domínguez, H., Parajó, J.C., Antioxidant properties of ultrafiltration-recovered soy protein fractions from industrial effluents and their hydrolysates. *Process Biochem.*, 41, 447–456, 2006.
88. Jiménez-Escrig, A., Alaiz, M., Vioque, J., Rupérez, P., Health-promoting activities of ultra-filtered okara protein hydrolysates released by in vitro gastrointestinal digestion: identification of active peptide from soybean lipoxygenase. *Eur. Food Res. Technol.*, 230 (4), 655–663, 2010.
89. Tsuruki, T., Kishi, K., Takahashi, M., Tanaka, M., Matsukawa, T., Yoshikawa, M., Soymetide, an immunostimulating peptide derived from soybean β-conglycinin, is an fMLP agonist. *FEBS Lett.*, 540, 206–210, 2003.
90. Yoshikawa, M., Tsuruki, T., Soy and Breast Cancer Prevention. In: *Soy in health and disease prevention*, CRC Press, pp. 225–233, 2005.
91. Kitts, D.D., Weiler, K., Bioactive proteins and peptides from food sources. *Curr. Pharm. Des.*, 9, 1309–1323, 2003.
92. Wu, J., Ding, X., Characterization of inhibition and stability of soy-protein-derived angiotensin I-converting enzyme inhibitory peptides. *Food Res. Int.*, 35, 367–375, 2002.
93. Shin, Z.I, Yu, R., Park, S.A., Chung, D.K., Ahn, C.W., Nam, H.S., Kim, K.S., Lee, H.J., His-His-Leu, an angiotensin I converting enzyme inhibitory peptide derived from Korean soybean paste, exerts antihypertensive activity in vivo. *J. Agric. Food Chem.*, 49, 3004–3009, 2001.
94. Gu, Y., Wu, J., LC–MS/MS coupled with QSAR modeling in characterising of angiotensin I-converting enzyme inhibitory peptides from soybean proteins. *Food Chem.*, 141, 2682–2690, 2013.
95. Ohinata, K., Agui, S., Yoshikawa, M., Soymorphins, novel μ opioid peptides derived from soy β-conglycinin β-subunit, have anxiolytic activities. *Biosci. Biotechnol. Biochem.*, 71, 2618–2621, 2007.
96. Carpino, L.A., Ghassemi, S., Ionescu, D., Ismail, M., Sadat-Aalaee, D., Truran, G.A., Mansour, E.M.E., Siwruk, G.A., Eynon, J.S., Morgan, B., Rapid, continuous solution-phase peptide synthesis: application to peptides of pharmaceutical interest. *Org. Process Res. Develop.*, 7, 28–37, 2003.
97. Nilsson, B.L., Soellner, M.B., Raines, R.T., Chemical synthesis of proteins. *Ann. Rev. Biophys. Biomol. Struc.*, 34, 91–118, 2005.
98. Yang, B., Mei, Y., Wang, X., Deng, X., Qian, H., Huang, W., Design, synthesis and biological evaluation of novel peptide MC62 analogues as potential antihyperglycemic agents. *Eur. J. Med. Chem.*, 73, 105–111, 2014.

99. Abraham, T., Prenner, E.J., Lewis, R.N.A.H., Mant, C.T., Keller, S., Hodges, R.S., McElhaney, R.N., Structure–activity relationships of the antimicrobial peptide gramicidin S and its analogs: aqueous solubility, self-association, conformation, antimicrobial activity and interaction with model lipid membranes. *BBA-Biomembranes,* 2014, 1838, 1420–1429.
100. Correa, W., Manrique-Moreno, M., Patiño, E., Peláez-Jaramillo, C., Kaconis, Y., Gutsmann, T., Garidel, P., Heinbockel, L., Brandenburg, K., Galleria mellonella native and analogue peptides Gm1 and ΔGm1. I) Biophysical characterization of the interaction mechanisms with bacterial model membranes. *BBA-Biomembranes,* 1838, 2728–2738, 2014.
101. Kim, Y.K., Yoon, S., Yu, D.Y., Lonnerdal, B., Chung, B.H., Novel angiotensin-I-converting enzyme inhibitory peptides derived from recombinant human α s1-casein expressed in Escherichia coli. *J. Dairy Res.*, 66, 431–439, 1999.
102. Prak, K., Maruyama, Y., Maruyama, N., Utsumi, S., Design of genetically modified soybean proglycinin A1aB1b with multiple copies of bioactive peptide sequences. *Peptides,* 27, 1179–1186, 2006.
103. Aleinein, R., Hamoud, R., Schäfer, H., Wink, M., Molecular cloning and expression of ranalexin, a bioactive antimicrobial peptide from Rana catesbeiana in Escherichia coli and assessments of its biological activities. *Appl. Microbiol. Biotechnol.*, 97, 3535–3543, 2013.
104. Losurdo, L., Quintieri, L., Caputo, L., Gallerani, R., Mayo, B., De Leo, F., Cloning and expression of synthetic genes encoding angiotensin-I converting enzyme (ACE)-inhibitory bioactive peptides in Bifidobacterium pseudocatenulatum. *FEMS Microbiol. Lett. (Wiley-Blackwell),* 340, 24–32, 2013.
105. Almaraz-Delgado, A.L., Flores-Uribe, J., Pérez-España, V.H., Salgado-Manjarrez, E., Badillo-Corona, J.A., Production of therapeutic proteins in the chloroplast of Chlamydomonas reinhardtii. *AMB Express,* 4, 57, 2014.
106. Gobbetti, M., Ferranti, P., Smacchi, E., Goffredi, F., Addeo, F., Production of Angiotensin-I-converting-enzyme-inhibitory peptides in fermented milks started by Lactobacillus delbrueckiisubsp. bulgaricus SS1 and Lactococcus lactissubsp. cremoris FT4. *Appl. Environ. Microbiol.*, 66, 3898–3904, 2000.
107. Álvarez-Martín, P., Belén Flórez, A., Margolles, A., del Solar, G., Mayo, B., Improved cloning vectors for bifidobacteria, based on the Bifidobacterium catenulatum pBC1 replicon. *Appl. Environ. Microbiol.*, 74, 4656–4665, 2008.
108. Álvarez-Martín, P., Flórez, A.B., Mayo, B., Screening for plasmids among human bifidobacteria species: sequencing and analysis of pBC1 from Bifidobacterium catenulatum L48. *Plasmid,* 57, 165–174, 2007.
109. Glick, B.R., *Molecular biotechnology: principles and applications of recombinant DNA,* 4th edn., ASM Press, Washington, DC, 2010.
110. Maliga, P., Bock, R., Plastid biotechnology: food, fuel, and medicine for the 21st century. *Plant Physiol.*, 155, 1501–1510, 2011.
111. Bock, R., Genetic engineering of the chloroplast: novel tools and new applications. *Curr. Opin. Biotechnol.*, 26, 7–13, 2014.
112. Matsui, T., Matsufuji, H., Seki, E., Osajima, K., Nakashima, M., Osajima, Y., Inhibition of angiotensin I-converting enzyme by Bacillus licheniformis

alkaline protease hydrolyzates derived from sardine muscle. *Biosci. Biotechnol. Biochem.*, 57, 922–925, 1993.
113. Kinsella, J.E., Melachouris, N., Functional properties of proteins in foods: a survey. *CRC Crit. Rev. Food Sci. Nutr.*, 7, 219, 1976.
114. Flambard, B., Johansen, E., Developing a fanctional dairy products: from research on Lactobacillus helveticus to industrial application of Cardi-04TM in novel anti-hypertensive drinking yogurt. In: *Functional dairy products*, M. Saarela, Ed., CRC Press, New York, vol. 2, pp. 506–518, 2007.
115. Wang, H., Zhang, S., Sun, Y., Dai, Y., ACE-inhibitory peptide isolated from fermented soybean meal as functional food. *Int. J. Food Eng.*, 9(1), 1–8, 2013.
116. Wilcox, S., The New Antimicrobial: Cationic Peptides. *Bio. Teach. J.*, 2, 88–91, 2004.
117. Chan, J.C.K., Li-Chan, E.C.Y., In: *Nutraceutical proteins and peptides in health and diseases*, Y. Mine, F. Shahidi, Eds., Taylor & Francis Group, New York, 2006.
118. De Vry, C.G., Valdez, M., Lazarov, M., Muhr, E., Buelow, R., Fong, T., Iyer, S., Topical application of a novel immunomodulatory peptide, RDP58, reduces skin inflammation in the phorbol ester-induced dermatitis model. *J. Invest. Dermatol.*, 125, 473–481, 2005.

8

Soy Protein Isolate-Based Films

Shifeng Zhang* **and Haijiao Kang**

College of Materials Science and Technology, Beijing Forestry University, Beijing, China

Abstract
This chapter discusses various topics of soy protein isolate-based films including, soy protein film preparation, characterization of soy protein films, modifications, and applications.

Keywords: Soy protein isolate-based films, soy protein film preparation, characterization of soy protein films

8.1 Introduction

Petroleum-based polymers complicate the processing of municipal waste due to their low weight-to-volume ratio and resistance to environmental degradation, and, as such, have become a rather serious problem [1–3]. Nondegradable petroleum-derived plastic materials, such as ethylene vinyl alcohol (EVA), polyethylene (PE), polyvinyl chloride (PVC), and polystyrene (PS), are currently quite commonly used in a variety of industries [4]. These materials are neither renewable nor biodegradable, generate excessive heat and gas when burned, and pose a serious threat to the environment. It has been predicted that the price of petroleum will rise considerably in future due to the excessive consumption of these types of polymers [5].

Most research on natural polymer materials focuses on systems which mimic the structures or functions of nondegradable synthetic polymer to

**Corresponding author*: zhangshifeng2013@126.com

Visakh P. M and Olga Nazarenko (eds.) Soy Protein-Based Blends, Composites and Nanocomposites, (195–230) 2017 © Scrivener Publishing LLC

some extent [6]. Natural polymers do indeed show many advantages, such as ready availability, complete biodegradation, high soil fertility, low accumulation in the environment, and successful waste management performance [7]. The ideal biodegradable material shows balanced mechanical and barrier properties as well as biodegradability. At least one step in the degradation process of a biopolymer is the metabolism of natural organisms, typically through composting or feedstock recycling. It has been reported that protein-based materials can fully degraded in 50 days when buried in soils [8].

Generally, naturally polymeric materials can be split into three categories according to the manner in which they are acquired: (1) plant-based polymers such as starch [9], soy protein [2, 3, 10], and cellulose [11]; (2) animal-based polymers such as chitosan [12] and keratin silk [13, 14]; and (3) microorganism fermentation-based polymers such as polyhydroxyalkanoates (PHAs) and polyhydroxybutyrates (PHBs) [15]. The biopolymers produced by organic molecules also include polyurethane (PU) from vegetable oil polyols, and PLA from completely renewable lactide monomers. In addition, it is worth noting that certain petrochemical-based polymers are biodegradable, including polybutylene succinate (PBS), poly(e-caprolactone) (PCL), and polyvinyl alcohol (PVA) [16]. Any of the above biopolymers contain polysaccharides and protein that are sorted, producing film that forms an effective barrier against oxygen (O_2), carbon dioxide (CO_2), and low-polarity compounds. Polysaccharides are composed of few or one monomer, while proteins are comprised of several amino acids [17].

Proteins are comprised of 20 different amino acids (lysine, leucine, phenylalanine, tyrosine, aspartic and glutamic acid, etc.) joined by peptide bonds [18], which spiral into three-dimensional conformation and fold into more complex structures; the primary structure is a linear polymer with a string of amino acids coupled by peptide bonds, and along the polymer chain is where the protein assumes different structures. The geometry of a protein is arranged in α-helix, β-sheet, or turn structures, which are coiled and stabilized through van der Waals, hydrogen bond, electrostatic, hydrophobic, and disulfide interactions. The tertiary structure is the overall shape of a protein molecule, which is stabilized by intermolecular interactions of lateral chains involving the series of hydrophobic amino acid residues and disulfide bonds formed among cysteines. The fewer the disulfide bonds, the weaker the rigid structures, while stronger, stiffer, and harder structures are formed by larger quantity of disulfide bonds. Intrinsic protein structure controls interaction behavior between these and any other composite components [19]. The unique structures of proteins allow them a wide range of functions; intra- or intermolecular binding gives a protein the potential to form bonds at diverse positions. In short, the mechanical properties of protein-based films are better than those of polysaccharide or fat-based films [20]. Compared to most plant protein-based films, soy protein isolate (SPI) films are clearer, smoother, and more flexible; they also show better gas barrier properties than films made by lipids or polysaccharides [21–23].

As mentioned above, fully sustainable and eco-friendly materials have become an urgent issue for researchers and developers in the polymer material industry [24–27]. Among all available biopolymers, animal and vegetable proteins are the most practical for making food packaging films due to their relatively low cost and high availability as by-products of the food processing itself.

A by-product of the soy oil extraction process, SPI is a highly refined or purified form of soy protein – over 90% of its protein content is obtained from defatted soy flour. SPI has higher protein content and better film-forming ability than other SP products such as soy flour (SF) which has 54% protein, or soy protein concentrate (SPC), which has 65–72% protein [2, 3, 28]. Based on its molecular weight and sedimentation coefficient, soy protein can be separated into 2S, 7S, 11S, and 15S fractions [29]. The 7S and 11S fractions account for 35% and 52%, respectively, of the soy protein content, where globular proteins β-conglycinin (7S) and glycinin (11S) are the two major protein components [30]. Both are tightly linked via hydrophobic, hydrogen bonding, and disulphide bonds in their subunits. The functional characteristics of soy protein are closely related to its physical and chemical properties, for example, molecular weight, composition and amino acid sequence, spatial structures, and surface charge. As such, SPI shows considerable potential for use in food, agriculture, bioscience, and biotechnology industry applications. That said, several problems limit their application in practice, including relatively poor mechanical properties, and poor water resistance and high moisture sensibility compared to petroleum-based plastics. These limitations due to the inherent hydrophilicity of the material as well as the strong molecular interactions of natural proteins. An effective cross-linking process is necessary to prepare films with acceptable mechanical and barrier properties while maintaining biodegradability. Various modification methods (including bulk and surface modifications, and physical and chemical modifications) have been developed to improve SPI-based materials, with mixed results.

Generally, physical modification primarily focuses on directional deformation. Hydrothermal treatment is one of the commonly applied physical modification methods for SPI [31]. Under high temperature and high-speed stirring conditions, the disulfide bonds among protein chains are unfolded to generate sulfhydryl groups. The resulting rearrangement of protein molecular spatial structures ensures favorable dispersion, emulsification, and solubility properties in the material, as well. Chemical modification of soy proteins, including cross-linking, acetoxylation, phosphorylation, glycosylation, and thiolation, targets active groups on the protein such as amino, hydroxyl, carboxyl, and sulfydryl groups [18]. The present study focuses on chemical modification.

Scheme 8.1 shows the reaction between acetic or succinic anhydride with soy protein. The amino groups in the soy protein and the carbonyl group embedded in the acetoxylation reagent react with each other and introduce hydrophilic groups onto the protein chains [32]. Enzymatic modification is an another efficient way of modifying protein – compared to cross-linking by chemcial reagents, enzymes can safely catalyze the covalent cross-linking of protein molecules, lending high molecular weight. After treating SPI with horseradish peroxidase, for example, the mechanical and water resistance properties of the SPI film significantly increase [33].

8.2 Soy Protein Film Preparation

Many previous researchers have proposed methods of enhancing SPI-based film performance, including custom processing methods [34], cross-linking with chemicals like formaldehyde [35], glyoxal [36], and glutaraldehyde (GA) [37,38], phenolic compounds [8, 39], and epoxy compounds [2, 3, 10, 40], polymerizing with functional groups [41], and

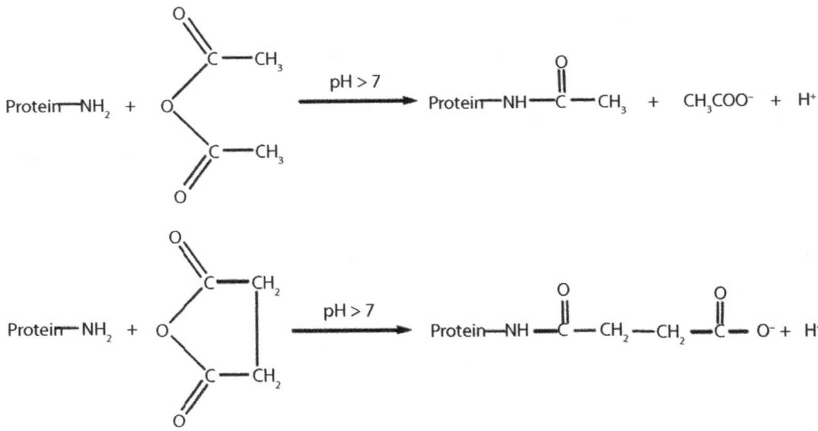

Scheme 8.1 Reaction of acetic and succinic anhydride with soy protein [32].

blending with PGA [42], PVA [43], PCL [44], PLA [45], PU [46], chitin [47], wheat gluten [48], gelatin [49], keratin [50], milk protein [51], zein [52], or silk [53].

Protein-based film preparation is typically conducted in three steps: (1) relieve the low-energy intermolecular bonds; (2) rearrange and orient the protein chains; and (3) form three-dimensional networks stabilized by the new linkages and bonds [54]. The most common methods used to fabricate SPI films are wet or dry processes. The wet process is based on dispersion or solubilization of proteins, where film mechanical properties are influenced significantly by the extrinsic conditions during film formation (such as air temperature and relative humidity) [28,55]. The dry process is based on thermoplastic protein processing, and is also used to prepare protein-based films [30, 46, 56].

8.2.1 Wet Process

The wet process, also called "solution casting," is based on solvent evaporation in the dispersion or solubilization of proteins. To this effect, the solvent components and pH of the solution are crucial to successful film formation [57]. Soy protein solubility is a function of the amino acid composition and sequence, and solvent pH value near or far from its inherent isoelectric point affects solubility and chain unfolding behavior [58]. Far from the isoelectric point (4.2–4.6 for SPI), the solubility of protein increases and its chains grow outstretched [8,23]. The solvent medium usually has elevated pH (alkaline, pH > 9) and prethermo-treatment is applied to stretch the native protein chains before pouring the material onto a flat surface [31,59]. Parameters of the continuous casting process including air temperature, surface properties of the formed substrate, flow rate, and drying time must be optimized appropriately [60]. The solvent drying step can be conducted under hot air, infrared energy, or microwave energy [61]. The solution casting process is generally slow and requires large volumes of solvents or water, so it is not often applied in the commercial field [62].

8.2.2 Dry Process

The dry process, also called "film extrusion," is a relatively more conventional approach than the wet process, and allows films to be obtained with a thermomechanically shaped extrusion machine [63]. During this process, protein chains are reassembled in an oriented manner to form molecular networks through unfolding, recombining, and cross-linking [64, 65]. The protein component forms networks through disulfide bonds or amide bonds between free carboxyl and amino side groups on the protein chains. Extrusion parameters such as mechanical energy input, shear stress level, pressure, and plasticizers, are important when preparing synthetic or biodegradable polymers. The plasticizer is particularly indispensable as far as protein bulk because without it, the protein molecules degrade, fixing the protein chain movement and increasing the viscosity of the system beyond what is acceptable [66].

Film extrusion is often utilized in packaging or other related fields, and has proven economically beneficial [28]. Generally, the extrusion process contains three steps: (1) feeding raw materials into a hopper; (2) conveying them by screw system; and (3) pushing them out in the desired shape [67]. In the studies of Zhang *et al.*, a multistep process was used to form SPI sheets by mixing SPI, water, glycerol, and other additives in a high-speed mixer, then applying a twin-screw extrusion to form pellet-shaped material [68].

The pelletized mixtures were then extruded into sheets using a single-screw extruder operated at 20–25 rpm. Barrel temperature varied from 120 °C to 160 °C. Pellets and sheets were shaped after cooling, and protein/glycerol mixtures were extruded using a twin-screw extruder with a screw speed of 100 rpm [26]. Table 8.1 shows the optimum extrusion temperature (T_E) in their studies for basic or acidic protein, the measured melt temperature (T_M), and the protein/glycerol ratios.

Compression molding can be applied to produce protein-based films of various types with varying extent of physical and chemical modifications during thermoplastic processing. This allows the thermoplastic and mechanical properties of the resultant films to be readily analyzed at a laboratory scale as a precursor to continuous extrusion to determine the most suitable processing conditions for the material [69]. Processing parameters including temperature, pressure, pressing times, and moisture content are responsible for the deformation of mixtures into viscoelastic

Table 8.1 Optimum extrusion conditions for protein/glycerol mixtures.

	Temperatures (°C)			Component ratio (w/w)	
Protein	**Extruder (TE)**	**Die (TD)**	**Melt (TM)**	**Protien isolate**	**Glycerol**
Basic soy protien	145	135	144	45	55
Acidic soy protein	150	135	138	45	55

melts. Intra- and intermolecular reactions of hydrogen, ions, hydrophobic, or covalence bonds then form during the cooling process. Ciannamea *et al.* prepared SPI films from SPC and glycerol by compression molding and solution-casting methods, and found that the formation process plays a major role in determining the properties of the resultant films. Compared to films prepared by casting, those prepared by compression mold had higher tensile strength (TS) and elastic modulus (E) as well as greater O_2 permeability and water resistance [30]. Guerrero *et al.* studied SPI films plasticized by glycerol through compression methods, and found that pH value, heat, and pressure significantly impact the performance of the resultant films; due to the effects of protein unfolding, films prepared at basic pH showed higher TS [70].

8.3 Characterization of Soy Protein Films

8.3.1 Mechanical Properties

In order to determine the performance of the resultant SPI films, they were cut into strips of specific sizes and mounted between the grips of the universal machine at set speed. The stress-strain curves were recorded by computer, as well as the mechanical properties of the soy protein films, including TS, ε_b, and E. "TS" was defined as the maximum TS a material can sustain, and was determined according to the maximum load exerted on the test specimen during the procedure [27]. Consequently, ε_b was measured at the point where the specimen broke, and expressed as the percentage of change compared to the length of the original material [71]. E was defined as the slope of the stress-strain curve in the elastic deformation region.

Mechanical properties are dependent primarily on composition and environmental factors, as well as processing conditions. For example, the plasticizer may increase polymer chains movement, causing ε_b and TS to decrease [65,72]. In a study by Guerrero *et al.*, different ratios of glycerol were employed to improve the mechanical properties of SPI films; glycerol ratio range of 30–40 wt% showed favorable plasticizing effect [73].

Cross-linking agents such as GA [37, 38, 74], epichlorohydrins [75], isocyanate [33,76], and genipin [77] are commonly used to improve the mechanical properties of polymer materials. Recently, phenolic compounds have also been used as cross-linking agents to enhance the performance of SPI films. In a study by Friesen, both rutin and epicatechin (Figure 8.1a) were shown to enhance the TS of SPI films from 9.3 MPa to 35.1 MPa and 22.1 MPa, respectively, as shown in Figure 8.1b [8]. Due to the cross-linking structure

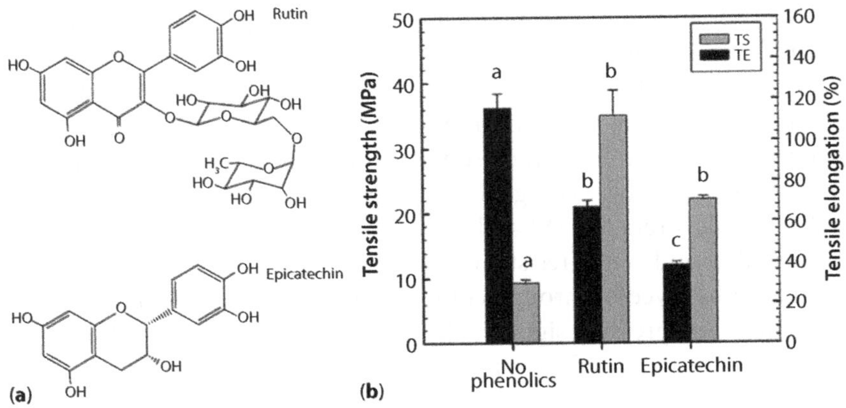

Figure 8.1 Structure of rutin and epicatechin (a) and TS and tensile elongation of SPI films with or without rutin and epicatechin (b) [8].

Table 8.2 TS, strain at break (ε), *E*, and puncture tests of films at pH 2, 8, and 11 [78].

	Tensile tests			Puncture tests	
Sample	**TS (MPa)**	**ε (%)**	**E (MPa)**	f_b **(N/mm)**	d_b **(mm)**
Film pH 2	1.50 ± 0.09 a	92.64 ± 13.72 a	0.292 ± 0.014 a	37.05 ± 2.50 a	2.17 ± 0.08 a
Film pH 8	0.45 ± 0.05 b	163.59 ± 14.91 b	0.029 ±0.007 b	13.82 ± 1.33 b	4.12 ± 0.57 b
Film pH 11	1.90 ± 0.28 c	205.41 ± 25.61 c	0.161 ± 0.019 c	40.22 ± 6.19 c	3.97 + 0.16 c

Values are means ± SD. Means followed by the same letter are not significantly different (P < 0.05) according to Tukey's test.

formation, the ε_b of the films sharply decreased. Mauri *et al.* investigated changes in solubility and molecular structure properties of SPI films prepared at different pH values (2,8,11). Their results showed that films obtained at pH 2 and 11 had denser microstructures and more disulfide bonds than those at pH 8, resulting in better TS and higher *E* (Table 8.2) [78].

Environmental humidity is another important factor that affects the mechanical properties of polymer films. Hydrophilic films more readily to uptake moisture at relatively higher humidity levels than hydrophobic films, and as such, also act as a sort of plasticizer to reduce the film's TS and enhance its extendability [79]. The specific processing method employed to obtain the film also influences its Films processed by compression show much better mechanical properties, higher TS, and elongation at the break(ε) than those processed by casting methods [73].

8.3.2 Dynamic Mechanical Analysis

Dynamic mechanical analysis (DMA) is becoming more and more commonly used in laboratory settings; it facilitates rheological analysis, which typically focuses on the deformation and flow of materials [80]. DMA provides information regarding major transitions as well as secondary and tertiary transitions in materials that are not readily identifiable by other methods [81]. DMA is based on fundamentally different responses of viscous and elastic elements at controlled temperatures, thus detecting and measuring the temperatures at which the amorphous domains of proteins reach free-flow state. Moreover, DMA is capable of detecting glass transitions according to the breadth of the transition or changes in heat capacity, which is not possible by differential scanning colorimetry (DSC) analysis [82]. The differences in T_g can be attributed to structural variations, molecular weights, and intermolecular interactions of different materials [83]. During SPI film preparation, changes in rheological properties due to plasticization can be predicted by DMA analysis, allowing more efficient and effective process design and resulting in better-performing films.

In a study by Mo *et al.*, the thermal properties of molded SPI films plasticized by four polyol-based plasticizers were analyzed using DMA methods ranging from −30 °C to 200 °C at a rate of 5 °C/min, under a static force of 660 mN and dynamic force of 600 mN at 1 Hz frequency. As shown in Figure 8.2, the storage modulus (E′) decreased with increasing temperature, and the onset temperature was the critical point where E′began to drop. T_g, an important property of amorphous polymers, can be identified at the peak temperature of loss modulus E″. In a broad temperature range, such as that of the unplasticized SPI sample shown in Figure 8.2, T_g may be dependent on the variety of peptide chains in the SPI [84].

8.3.3 Thermogravimetric Analysis

Thermogravimetric analysis (TGA) is a useful analysis technique for studying the thermal stability of polymers. TGA is typically employed according to measured weight changes in a sample as it is heated, cooled, or held at a constant temperature.

Guo *et al.* investigated the thermal stability of SPI/polybutylene adipate-*co*-terephthalate (PBAT) blends with a thermo-gravimetric analyzer in which samples (5–10mg) were heated from 25 °C to 600 °C at a rate of 10 °C/min under N_2 atmosphere conditions. Figure 8.3 shows where the pure

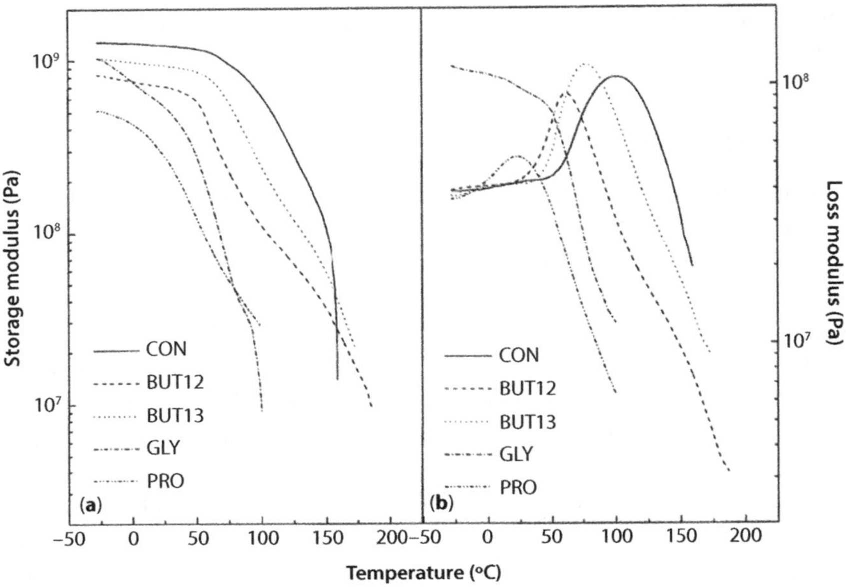

Figure 8.2 DMA of E′ (a) and loss modulus (b) on temperatures for SPI plastics.
Note: BUT12, BUT13, GLY, PRO, and CON represent SPI plastics with 1,2-butanediol, 1,3-butanediol, glycerol, propylene glycol, and unplasticized SPI plastics, respectively [84].

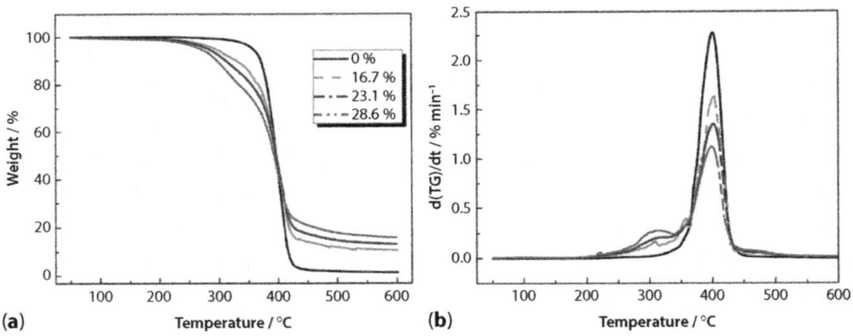

Figure 8.3 TGA (a) and DTG (b) curves of SPI/PBAT blends in N_2 atmosphere [85].

PBAT was stable until 350 °C, and the peak of 400 °C in differential thermal gravity (DTG) curve reflects its decomposition temperature. The SPI/PBAT blends were all stable below 200 °C, and a two-step weight loss was observed between 200 °C and 400 °C in line with the two DTG curve peaks around 220 °C and 310 °C, owing to the processing additives and SPI degradation, respectively. The addition of SPI improved the material's thermo-stability, evidenced by increased residue weight after thermo-degradation [85]. In a study by Ogale *et al.*, TGA was used to test the thermal degradation of SPI powder and SPI-glycerol films. The SPI powder and plasticized films were thermo-stable before 180 °C and 150 °C, and significantly degraded above 200 °C and 180 °C, respectively. The optimum compression-molding temperature of SPI-glycerol mixtures was identified at 150 °C [80].

8.3.4 Differential Scanning Colorimetry

Differential scanning colorimetry (DSC) measures the difference in heat flow rate between samples subjected to a controlled temperature [68,86]. DSC is often used to characterize the thermal transitions of polymers and protein-based films by detecting T*g*, melting processing, crystallization, and degradation behavior. The plasticizing effect of plasticizers decreases the T*g* value of polymer sheets, as confirmed by DSC analysis. The characteristic T*g* alternation identified by DSC confirms protein denaturation or molecule reassembly during processing [69].

Di Gioia and Guilbert used modulated-DSC to study the plasticization performance of water, glycerol, and tanoic acid, and found that T*g* gradually decreases as plasticizer concentration increases [87]. The thermo properties of soy protein solutions and the soy protein films were both investigated in Adriana's research [31], where, as shown in Figure 8.4a, two typical endotherms were identified at 78.2 ± 0.2 °C and 92.4 ± 0.1 °C at pH 8, corresponding to β-conglycinin and glycinin denaturations, respectively. In addition to acid treatment, both glycinin and β-conglycinin treatment can affect the extent of denaturation, decreasing enthalpy. Thermal treatment at 60 °C does not affect the folding of proteins at pH 8, but SPI at pH 2 completely denatures and at pH 11, glycinin partly maintains its original structure (Figure 8.4b).

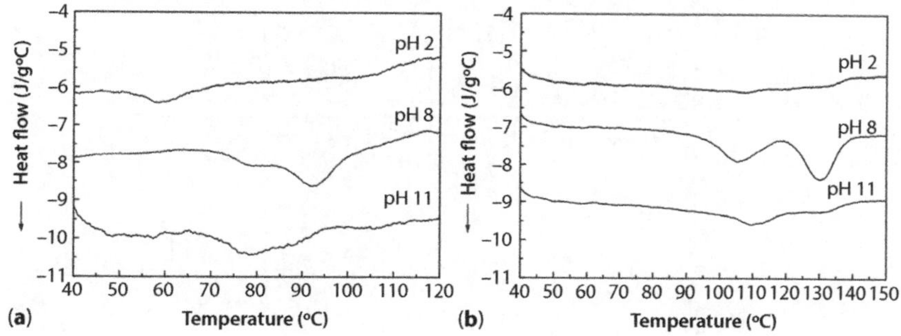

Figure 8.4 DSC of SPI solutions (a) SPI films (b) at pH 2, 8, and 11 [31].

8.3.5 Scanning Electron Microscopy and Transmission Electron Microscopy

Scanning electron microscopy (SEM) and transmission electron microscopy (TEM) are very often utilized to determine the morphological manifestation of known biochemical changes that accompany development [88]. SEM, which replaces light sources with high-energy electron beams, is one of the most versatile instruments available for examination and analysis of microstructural morphology and chemical composition of any material. Objects can be subtended to about a 1/60° visual angle, corresponding to a resolution of ~0.1 mm. SEM has been widely applied to observe the morphology of protein-based materials and to determine whether or not certain physical/chemical interactions [10].

During TEM analysis, an electrons beam is transmitted through an ultrathin specimen and interacts with the specimen as it passes through, then an image is formed from the interaction and magnified onto a fluorescent screen or layer of photographic film, or detected by a sensor. TEM provides images at higher resolution than light microscopes, enabling specimens to be examined in extremely fine detail – even down to single column of atoms. TEM is generally employed to investigate biopolymer crystallization, as well as visual assessment of particle size or morphology. Xia *et al.*, for example, used SEM to study the cross-sectional morphology of tensile fracture surfaces in SPI films, as shown in Figure 8.5. The results indicated that the modified SPI films had homogeneous surfaces with fewer defects and inhomogeneous characteristics compared to the unmodified SPI films.

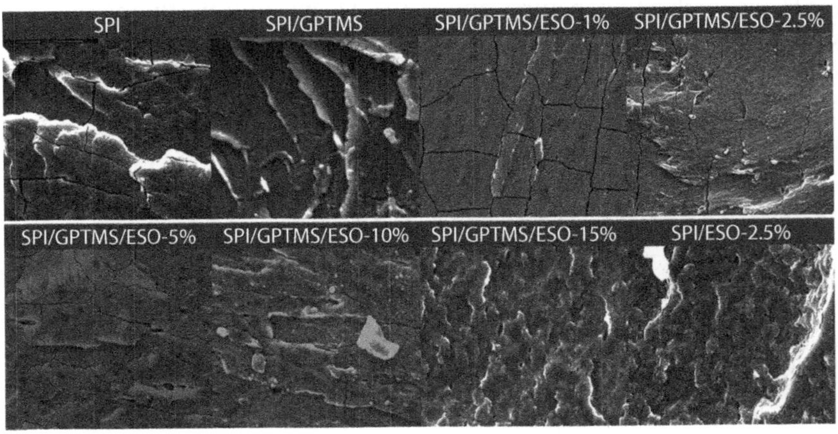

Figure 8.5 SEM images of cross-sections of the SPI-based films [10].

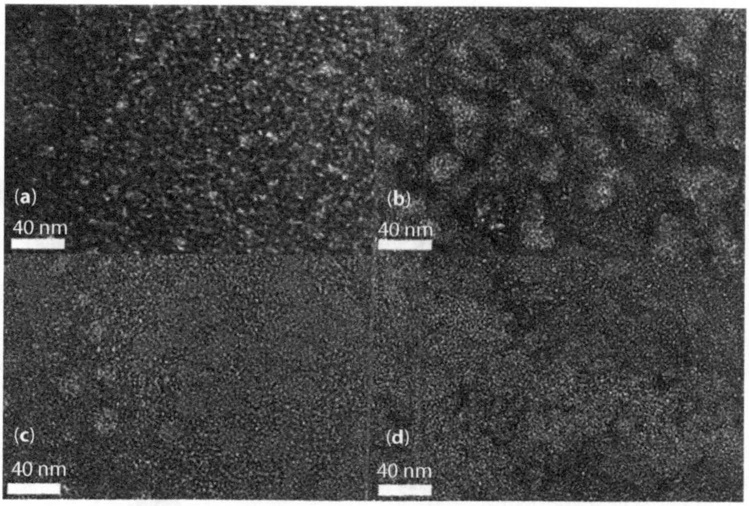

Figure 8.6 TEM images of SPI and modified SPI samples [41].

Other materials and processes can be analyzed effectively using TEM, for example, the hydrophobic surface created by grafting polymerization onto soy protein shown in Figure 8.6 [41].

8.3.6 Atomic Force Microscopy

Atomic force microscopy (AFM) uses a high-resolution instrument, acquiring analysis data by observing weak interatomic forces between the probe and the specimen surface. The AFM technique can be applied to measure the roughness of specimen surfaces and determine mechanical hardness, roughness, and micro-fabrication properties in the specimen, as shown in Figure 8.7.

8.3.7 Rheological Measurements

Proteins, which are inherently polymeric and polyelectrolytic, can stabilize films, foams, emulsions, and dispersions through a combination of steric and electrostatic (electrosteric) repulsion due to absorption at the interface between two different phases [90]. Thus, the stability of protein-based films under physical stresses caused by emulsions, foams, or dispersions plays a crucial role in the film-forming process, and is believed to be related to the interfacial dilatational and shear rheology of the adsorbed film. Interfacial rheology analysis techniques probe the interactions within adsorbed films,

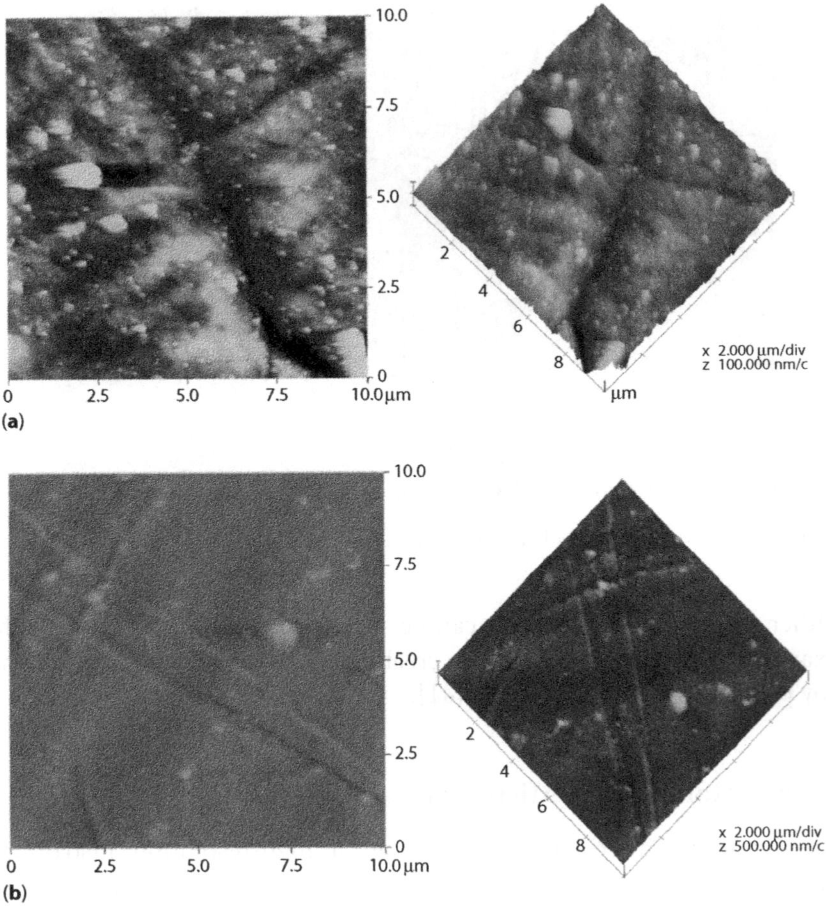

Figure 8.7 Surface morphology of SPI materials (a) unmodified (b) modified SPI films [89].

whether unmodified protein films or films made of proteins mixed with other surfactants or in complex formation with other materials. There have been many recent improvements to interfacial rheology measurement techniques for protein-based films, including distinguishing between dilatational and shear rheology. The dilatational technique involves inducing changes in the interfacial area while simultaneously measuring interfacial tension, and the shear technique involves inducing shear in the film without any change in area; many different arrangements are available for shear measurement.

Wang *et al.* studied the interfacial dilatational rheology of soy protein at an oil-water interface, using an optical contact angle meter to monitor and record the temporal evolution of surface pressure and surface dilatational parameters [91]. A drop of protein solution was delivered and allowed to stand for 3 h to achieve protein adsorption at the oil-water interface. Equation (1) was used to correlate the change in interfacial pressure (π) with time, and $\pi = \sigma^0 - \sigma$ (where σ is surface tension.) Equation (2) was used to describe the surface dilatational modulus derived from the change in interfacial tension (σ), resulting from a small change in surface area (A).

$$\pi = 2C_0KT\left(D\theta / 3.14\right)^{1/2} \tag{1}$$

$$E = \mathrm{d}\sigma / \left(\mathrm{d}A / A\right) = -\mathrm{d}\pi / \mathrm{d}\ln A = E_d + iE_v \tag{2}$$

It is commonly accepted that interfacial rheology indicates the structural state of proteins and macromolecule interactions. Figure 8.8 shows the evolution of surface dilatational modulus with π value in SPI-based specimens. E value increases as π lengthens, and the maximum *E* value decreases as SPI concentration increases. The multifaceted slopes tested revealed rearrangement of the adsorbed primary layer and/or multilayer formation at a surface pressure higher than the equilibrium surface pressure.

Sun *et al.* applied a rotational rheometer to measure the E′(*G*′) and loss modulus E″ (*G*″) at different time points during the film-formation process [92], and the effects of oscillatory shear frequency on the viscoelastic properties of different samples were evaluated according to Eq. (3)

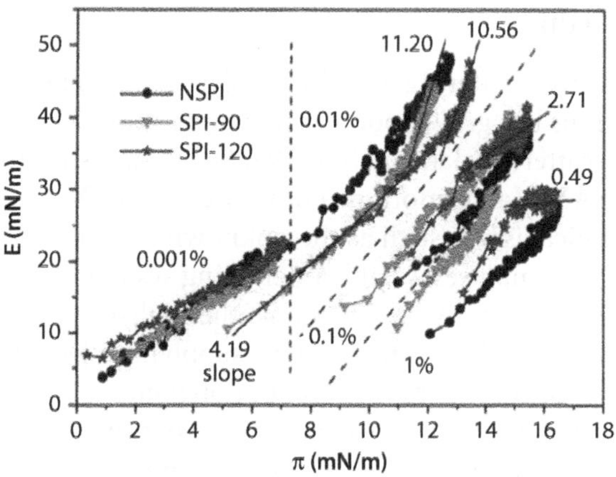

Figure 8.8 Surface dilatational modulus (*E*) as a function of surface pressure (π) for native and heat-treated SPI (SPI-90 and SPI-120) at the oil-water interface [91].

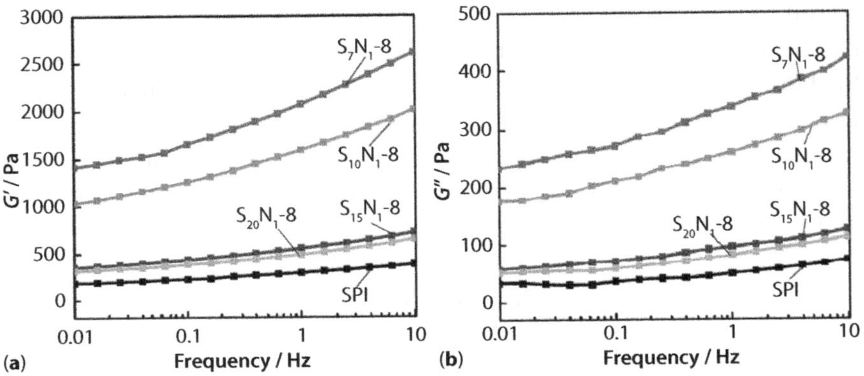

Figure 8.9 G'(a) and G''(b) curves of SPI and SPI-CNF to oscillatory shear frequency [92].

which was used to calculate the dynamic complex viscosity (η^*). As shown in Figure 8.9, compared to the control SPI film, SPI/cellulose nanofibril (CNF) composites showed higher G' and G'' values. SPI-CNF mixtures with smaller CNF proportions were frequency-independent, but other mixtures (S_7N_1,$S_{10}N_1$) were sensitive to frequency, proving that CNF effectively enhances SPI viscoelasticity.

$$\eta^* = \left[(G')^2 + (G'')^2 \right]^{1/2} / \omega \qquad (3)$$

8.3.8 Small-Angle X-Ray Scattering and Wide-Angle X-Ray Scattering

Small-angle X-ray scattering (SAXS) analysis of biological macromolecules in solutions has grown popular due to its automated data collection and in-depth scattering data analysis [93, 94]. Many structural biologists have applied SAXS (and, more recently, wide-angle X-ray scattering) to challenge biological issues such as the unknown flexibility of certain macromolecules and complexes, weakly interacting systems, and dynamic processes [95]. SAXS elegantly complements the high-resolution probes that may be limited by problematic samples, supplying valuable information about the shape, compactness, and size of polymers.

In a study by Wang, SAXS experiments were performed using a SAXSess camera equipped with a PW3830 X-ray generator and sealed glass X-ray tube. The intense monochromatic primary beam was provided by the focusing multilayer optics and block collimator. And a semitransparent beam stop enabled the measurements of the attenuated primary beam at the zero scattering vector ($q = 0$) [91]. The samples (10 mg/mL) were filled into vacuum-tight thin quartz capillaries and placed in sample holder unit, where test distance and room temperature were 261.2 mm and 26.0 °C. The two-dimensional scattered intensity distribution was then recorded by an imaging-plate detector, then the data were integrated into 1D scattering function $I(q)$ as a function of the magnitude of scattering vector q ($q = 4\pi \sin \theta/\lambda$, where 2θ is the scattering angle.) All $I(q)$ data were normalized according to the uniform primary intensity at $q = 0$. Samples of heat-treated soy proteins were probed by SAXS and the results are shown in Figure 8.10b. It can be concluded that the spherical morphology of native proteins is in line with the typical bell shape [96]. Meanwhile, the magnitude of maximum peak changes to be lower with increased treated temperature, indicating that higher temperature is beneficial to the more flexible conformation.

Generally, the Kratky plot of native soy protein displays a maximum peak based on the degree of compactness of its proteins, then a plateau, followed by monotonic increase reflecting unfolded polypeptides. Native SPI exhibits a typical bell shape confirming its spherical morphology; heat-treated SPI samples shows decreased magnitude of maximum peaks, suggesting the presence of globular shapes and partially unfolded structures of rigid glycinin and β-conglycinin. Based on the SAXS data, that is, ln $I(q)$ versus q^2, the intensity to an angle of zero $I(0)$ and the gyration radius of protein Rg were extrapolated as summarized in Table 8.3. $I(0)$ is proportional to molecular weight, while Rg

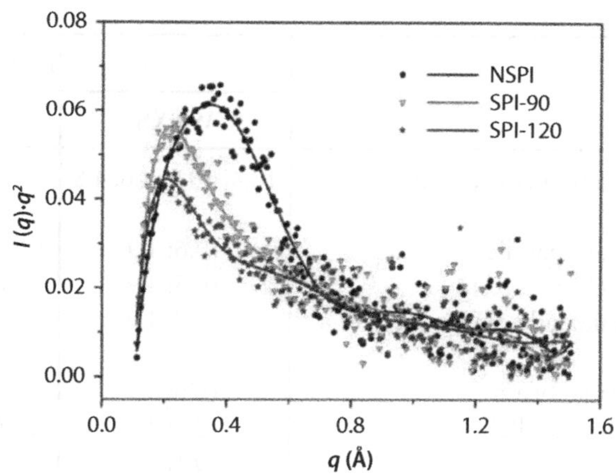

Figure 8.10 SAXS of SPI-based samples [91].

represents size or molecular compactness. The quantitative variations of *I(0)* and *Rg* value reveal that heat treatment results in more flexible protein conformation.

In the study by Lai *et al.*, both wide-angle and small-angle X-ray scattering (WAXS and SAXS) techniques were used to study dry granular zein-based materials [97]. The WAXS results for zein-oleic acid resin and films showed that granular zein and zein fibers have similar d-spacings (4.6 Å and 10.5 Å, 9.5 Å, respectively), as shown in Table 8.4. Changes in interhelix spacing were related to the preparation process, whereas the basic internal structure remained unchanged. Diffuse rings indicative of noncrystalline structures were also observed in the samples. SAXS analysis likewise confirmed that processing affected film morphology. The authors

Table 8.3 Structural parameters of native and heat-treated SPI derived from SAXS and DLS data.

Samples	*I*(0)[a]	R_g (nm)[a]	R_h (nm)[b]
NSPI	1.75 ± 0.02	5.70 ± 0.05	16.95 ± 0.16
SPI-90	2.89 ± 0.04	7.49 ± 0.04	17.69 ± 0.13
SPI-120	2.58 ± 0.04	7.91 ± 0.05	20.37 ± 0.06

[a] Guinier approximation of SAXS data; *I*(0), the extrapolation of intensity to an angle of zero; R_g, the gyration radius of protien. [b]From DLS experiments; R_h, the hydrodynamic radius of protien.

Table 8.4 Spacing values of zein-based materials tested by WAXS and SAXS [97].

Samples	d-Spacing (Å)	
	WAXS	SAXS
Resin film, beam normal, (biaxially deformed)	10.4, 10.9; 4.6, 4.8	165, ~45
Resin film, beam parallel	10.6; 4.9	135 (°); 170 (90°)
Dried zein–oleic acid resin	10.5; 4.9	165, ~45 (?)[a]
Resin film (uniaxially deformed with a roller)	10.7; 4.9	170
Cast film, normal beam	10.6; 4.9	80; 45 (?)
Cast film, beam parallel	10.2; 4.9	65 (0°); 80 (90°)
Granular zein	10.1; 4.7	–[b]
Zein fiber	9.9; 4.7	–[b]

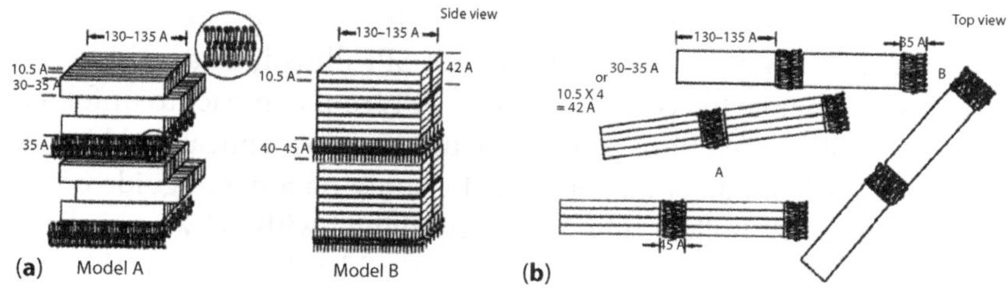

Figure 8.11 Proposed structure models of zein-oleic acid resin films, side view (a) and top view (b), aggregated longitudinally to form platelets [97].

also proposed several possible structural models of zein-oleic acid resin films based on the data they gathered (Figure 8.11).

8.4 Modifications

Due to its relative abundance and favorable film-forming performance, soy protein is often used to develop biodegradable membranes. Researchers have proposed physical, chemical, and combination modifications in effort to extend the practical application of SPI films and overcome disadvantages such as poor mechanical properties and moisture sensitivity [98,99].

Blending SPI with other biodegradable polymers is generally accepted to be successful [45,100]. Blending methods do show favorable reinforcing effects, and even low nanofiller loads produce better SPI-based bionanocomposites than SPI alone [101–103]. The development of new nanocomposites has garnered considerable interest as it may improve polymer properties and extend their practical application [104,105]. "Nanoclusters" (NCs) are defined as ultrafine, isolated particles less than 2 nm in size with several to a hundred atoms, which are located in the transition region between atoms and bulk solids [106,107]. Researchers have successfully produced a series of reinforced SPI sheets compounded with nano-SiO_2 particles in a variety of distributions (single nanosphere, ~100 nm nanocluster, interconnected network structure, and great domain,) in the SPI matrix. Results have shown that the simultaneous enhancement of strength and elongation is related to homogeneous dispersion of nanoclusters, while aggregated great domains severely decrease elongation despite significant reinforcing effect [108]. In addition, water-soluble Cu NCs and Zn NC-modified SPI films were synthesized by Li *et al.* to regulate the rigidity and flexibility of bio-films [109]. The specialized bio-films formed by SPI modified by Cu NCs exhibit better mechanical properties, and those modified with Zn NCs show better hydrophobic properties, than SPI alone. Further X-ray diffraction testing of SPI films indicated that metal nanoclusters change the conformation of SPI from compact to unfolded.

Modified SPI-based films can also be used in biochemical and biomedical research fields. Luo *et al.* conducted a study on a nerve guide conduit from cellulose/SPI hollow tube (CSC) combined with Schwann cells and pyrroloquinoline quinone, and found that the blended film has the ability to repair and reconstruct nerve structure and function owing to the comprehensive contributions from the hollow CSC tube [110]. The cytocompatibility and toxicity of the cellulose/SPI composite films (CSM) were also tested *in vitro*, and results showed that CSM had *in vitro* cytocompatibility with Schwann cells and was biologically safe for animals, suggesting potential application as a treatment for nerve defects in humans [111].

8.4.1 Physical Modification

Incorporation of filler materials in order to improve the barrier properties of polymers is a well-known technique in food packaging applications [112]. Owing to the availability of a wide variety of fillers, it is useful to distinguish between inorganic fillers, which may also be of natural origin, but as a general rule, do not consist of any carbon atoms (e.g., SiO_2, TiO_2, nanoclay.) Organic fillers, namely those containing carbon atoms, are most often derived from biomass (e.g., CNF.)

8.4.1.1 SiO_2

Wang H *et al.* investigated the effects of SiO_2 content, sodium dodecylsulfate, and electrospinning electrical field strength on the morphology and structure of PVA/SPI/SiO_2 hybrid fibers. Their results showed that the PVA, SPI, and SiO_2 nanoparticle components were linked by hydrogen bonds or Si-O-C bonds in the hybrid fabric networks. DSC tests showed that the presence of SiO_2 nanoparticles considerably improved the thermal stability of the hybrid fibers [113].

Wang SY *et al.* studied the effect of average particle size on modified SPI film performance [114]. Results showed that under two-hour UV irradiation at 365 nm, the bactericidal activity of SPI films containing TiO_2 nanoparticles (2.00 g/150 mL) against *Escherichia coli* and *Staphylococcus aureus* can reach 71.01% and 88.49%, respectively. Wang *et al.* also investigated the effects of nano-TiO_2 content and UMAT (ultrasonic/microwave assisted treatment) time on the physical properties and structure of films, and found that the intermolecular force between nano-TiO_2 and

SPI significantly enhanced the films' mechanical and barrier properties. UMAT time ≤20 min significantly improved films' TS values (15.77 MPa, 245% higher than the control,) and reduced water vapor permeability (1.8457×10^{-11} g cm^{-1} s^{-1} Pa^{-1}, 72.11% lower than the control) [115].

8.4.1.2 Nanoclay

In another study, Halloysite nanoclay and nano-SiO_2 were used as reinforcing materials in bovine gelatin based bio-nanocomposite films [116]. The composition included gelatin with 20% (w/w) of glycerol compounded with halloysite nanoclay and nano-SiO_2 (0%, 2%, 3%, 4%, 5% w/w), respectively. Both types of nanocomposite films showed better mechanical and water solubility than those of the control; between the two nanomaterials used, halloysite nanoclay showed better mechanical properties. These results support the concept of nanocomposite technology in general, and specifically suggest that NCs can be employed to effectively improve the barrier and mechanical properties of bovine gelatin films for food packaging purposes.

8.4.1.3 CNF

Lubasova *et al.* fabricated soy-protein-based CNF modified by green cross-linker-containing aldehyde (–CHO) groups [117]. Defatted soy flour was first purified through acid-washing to ensure high protein content, then blended with gluten, then electrospun into nanofibers assisted by polyvinyl alcohol. The green cross-linker was synthesized by sucrose oxidation with hydrogen peroxide (H_2O_2). Results showed that the water-resistant of protein-based nanofibers cross-linked by oxidized sugar are as good as those cross-linked using toxic glyoxal. In another related study, Jensen *et al.* prepared soy protein films incorporated with cellulose fibers that were extracted from soybean pods and stems using a combination of acid-alkaline hydrolysis and high-pressure homogenization treatment. For the sake of comparison, TiO_2-SPI composite films were also prepared by hot surface casting with TiO_2 nanoparticles as filler [118]. Mechanical and barrier properties (water vapor, and O_2 permeability) of these films were evaluated under different relative humidity (RH) conditions, and in general, TS and E decreased and ε_b increased as RH increased.

8.4.2 Chemical Modification

Chemical agents can be used to create intra- and intermolecular reactions that improve the mechanical strength, barrier properties, and thermostabilities of soy protein-based materials. Figure 8.12a shows covalent and

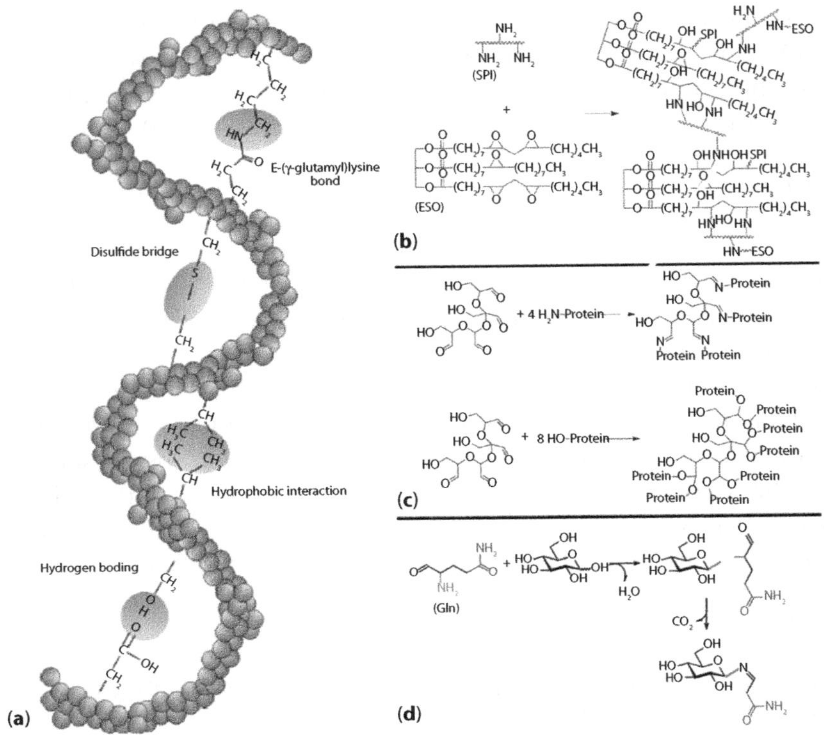

Figure 8.12 Schematic diagram of possible interactions in cross-linked protein films (a) [119,120]; cross-linking reaction mechanisms of SPI and ESO (b) [10]; oxidized sucrose and soy protein (c) [121]; and Maillard reaction between amino acids and CMC (d) [122].

non-covalent interactions formed after physical, chemical, and enzymatic treatments [119, 120] and Figure 8.12b–d shows the cross-linking reaction mechanisms of SPI and ESO [10] and between oxidized sucrose and soy protein, as well the Maillard reaction between amino acids and carboxymethyl cellulose (CMC).

8.4.2.1 *Aldehydes*

Aldehydes, including formaldehyde [35], glyoxal [36], GA [37, 38], and dialdehyde starch [123] are often utilized as cross-linking agents to modify SPI film. Formaldehyde is a common cross-linking agent that can interact with various amino acids like tryptophan, tyrosine, histidine, arginine, and cysteine. The TS of SPI films cross-linked by formaldehyde is twice more that of untreated films, and with lower water solubility (WS) and water vapor permeability (WVP) [124]. The side chain groups of histidine, cysteine, tyrosine, and lysine can be cross-linked by GA, which can also be utilized to cross-link the SPI. Cross-linked films show lower solubility and significantly higher TS than untreated films [125].

8.4.2.2 Isocyanate

Isocyanate is a type of unsaturation ester. Its residual –NCO groups can react with SPI functional groups to form urethane linkages. Methylene diphenyl diisocyanate (MDI) [33] and hexamethylene diisocyanate (HDI) [126] are the most commonly used cross-linkers, both of which contain two isocyanate groups and are highly reactive in forming both hydroxyl and carboxyl groups. Cross-linking structures improve the mechanical properties of the resultant composites significantly, but inevitably result in decreased elongation at the break [127].

Fang *et al.* prepared SPI/ PLA composites with or without MDI and found that TS of PLA/SPI material without MDI was very low due to the stress concentration on the dispersed SPI. After MDI was added, conversely, the TS of the composites significantly improved [128].

8.4.2.3 Epoxy

Based on the native and reactive groups in soy proteins like –OH, –SH, –COOH, and $-NH_2$, epoxy compounds such as ESO [10], epoxy resin [129], waterborne epoxy latex [130], and glycidyl methacrylate [99] can be used to enhance SPI film performances. A study by Xia *et al.*, for example, successfully used ESO for this purpose; results showed that after 2.5% ESO addition, the TS of the SPI film increased by 139.8% compared to the control.

8.5 Applications

As an abundant and renewable resource with many advantages, including renewability, biocompatibility, biodegradability, processability, and film-forming capacity, SPI shows attractive potential in food, medical, and agriculture applications.

8.5.1 Food Packaging

Due to the high polarity of SPI macromolecules, plus ordered hydrogen-bonded network structure and low-solubility, SPI films create semipermeable barriers to water vapor or oxygen between food and its ambient environment, effectively protecting foods from oxidative damage [131–134]. Films containing antimicrobials, such as organic acids, chitosan, or natural plant extracts, can be employed to delay the growth of contaminating microorganisms and to maintain food quality during storage [135].

Guerrero *et al.* found that fresh beef patties treated with SPI coating stayed fresh for 14 days as a result of SPI coating delaying the oxidation and deterioration of lipids. The surface color stability of the patties was also maintained during chilled storage [136]. In a similar study by Wang *et al.*, bio-nanocomposites of SPI/TiO_2 film caused a sizeable decline in water vapor and oxygen permeability (decreased by 72.12% and 57.64%, respectively,) compared to the control [115]. In a study by Usui *et al.*, soy protein-chitosan conjugate was synthesized through Maillard reaction and results showed that the antimicrobial activity of the soy protein/chitosan conjugates improved 2 to 3 times compared to a plain soy protein/chitosan mixture [137]. The allergenicity of soy protein can be significantly decreased by the attachment of chitosan through Maillard reaction, and soy protein-chitosan conjugate successfully blocks the allergen structure of soy protein that results from 34 kDa-protein (Gly m Bd 30K) compared to soy protein-galactomannan conjugate.

As techniques progress, the direct application of antibacterial substances on the surface of food still has its drawbacks. Current research on coating antibacterial substances onto biodegradable or edible films has shown mixed results [138]. In a study by Pan *et al.*, positively charged *N*-[(2-hydroxy-3-trimethylammonium) propyl] chitosan chloride (HTCC) and negatively charged SPI were alternately assembled on cellulose acetate (CA) electrospun nanofibrous mats via electrostatic layer-by-layer (LBL) self-assembly technique (Figure 8.13) [139], and results showed that HTCC and SPI did coat successfully on the surface of the CA mats and CA nanofibrous mats. Bilayers of HTCC/SPI possessed more orderly-arranged structure than uncoated ones. Further, the HTCC/SPI-coated nanofibrous mats showed excellent antibacterial properties against *Escherichia coli* and *Staphylococcus aureus* – the average diameters of inhibition zones were 9.6 mm and 11.53 mm, respectively.

8.5.2 Medicine

Protein-based nano-medicine platforms, including films, hydrogels, and micro- or nano-particles, are ideal materials for drug delivery [140]. Abundant active compounds on protein polypeptide chains can be embedded into these above-mentioned types of platforms, providing a wealth of possibilities for reversible binding of active molecules and controllable

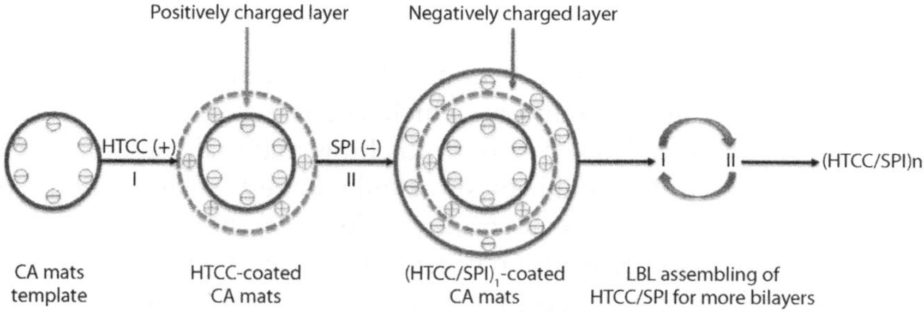

Figure 8.13 Scheme of LBL modification on mats [139].

release at the desired site *in vivo*. There has been extensive research conducted on animal protein-based materials, but newer soy protein-based films are infection-resistant and highly hydrophilic, and show favorable tensile strength, permeable properties, and controllable biodegradable features for film dissolution/swelling; as such, they represent a very innovative wave of medical applications. Soy protein-based materials have been explored for their ability to promote wound healing, and to replace disposable surgical gloves and infusion bags [134].

In recent study, protein-based materials for drug delivery systems including ferritin/apoferritin protein cages [141], plant-derived viral capsids [142], small Heat shock protein (sHsp) cages [143], elastin-like polypeptide (ELP) [144], and zein, soy, and whey protein, collagen, and gelatin [35,145–147], have been developed in designing the drug delivery system. The food protein in films serves a good platform for drug delivery system, which has the potential to interact with hydrophobic or hydrophilic drug molecules through its surface functional groups.

A study by Chen *et al.* examined the kinetics of two different drug delivery systems (hydrophilic/methylene blue and hydrophobic/rifampicin,) on SPI films in the gastric intestinal tract. Results showed that drug release was successfully controlled through erosion of the SPI film by zero-order kinetics [35]. The SPI film networks displayed excellent compound binding capacity, especially for hydrophobic molecules, and as such showed attractive potential for use in controlled release systems based on matrix erosion.

Innovative biodegradable wound dressings based on soy protein films were developed and studied by Peles *et al.* [148]; the films effectively inhibited *Staphylococcus aureus* and *Staphylococcus albus* infections for at least 2 weeks, and *Pseudomonas aeruginosa* for 3 days. The inhibition profiles obtained during 2 weeks of incubation fit the *in vitro* release profiles.

8.5.3 Agriculture

Plastic mulching cultivation technique has been a widely-spread method in agricultural industry, which could accelerate the immature plants growing. However, the mulching materials significatively depend on the non-degradable plastics, and release a sizeable amount of harmful substances into the environment [69,149]. Biodegradable film materials such as soy protein, however, can be directly introduced into the field soil or into the decomposition of microbes as an alternative [150, 151].

Starch-based biodegradable polymer has been utilized by agriculture industry workers in Spain for the past several years, and has proven environmentally friendly; its high cost and heavy weight are problematic [151,152]. Soybean material, conversely, is low in cost and more easily transported and processed than starch. That said, because soy protein contains more than half of the polar amino acids that cause hydrophilicity, it is difficult to eliminate the moisture sensitivity of SPI composites [153]. For this reason, as discussed above, physical and chemical (or alternative) modifications should be established for improving the properties of SPI. Research has already shown that doing so is well within our reach.

References

1. Ou, S., Wang, Y., Tang, S., Huang, C., Jackson, M.G., Role of ferulic acid in preparing edible films from soy protein isolate. *J. Food Eng.*, 70, 205–210, 2005.
2. Xu, F.J., Zhang, W., Zhang, S.F., Li, L., Li, J.Z., Zhang, Y., Preparation and characterization of poly(vinyl alcohol) and 1,2,3-propanetriol diglycidyl ether incorporated soy protein isolate-based films. *J. Appl. Polym. Sci.*, 132, 42578–42580, 2015.
3. Xu, F.J., Dong, Y.M., Zhang, W., Zhang, S.F., Li, L., Li, J.Z., Preparation of cross-linked soy protein isolate-based environmentally-friendly films enhanced by PTGE and PAM. *Ind. Crops Prod.*, 67, 373–380, 2015.
4. Hanani, Z.A.N., Roos, Y.H., Kerry, J.P., Use and application of gelatin as potential biodegradable packaging materials for food products. *Int. J. Biol. Macromol.*, 71, 94–102, 2014.
5. Song, F., Tang, D.L., Wang, X.L., Wang, Y.Z., Biodegradable soy protein isolate-based materials: a review. *Biomacromolecules*, 12, 3369–3380, 2011.
6. Schriemer, D.C., Li, L., Detection of high molecular weight narrow polydisperse polymers up to 1.5 million daltons by MALDI mass spectrometry. *Anal. Chem.*, 68, 2721–2725, 1996.

7. Li, Y.D., Zeng, J.B., Wang, X.L., Yang, K.K., Wang, Y.Z., Structure and properties of soy protein/poly (butylene succinate) blends with improved compatibility. *Biomacromolecules*, 9, 3157–64, 2008.
8. Friesen, K., Chang, C., Nickerson, M., Incorporation of phenolic compounds, rutin and epicatechin, into soy protein isolate films: mechanical, barrier and cross-linking properties. *Food Chem.*, 172, 18–23, 2015.
9. Lendvai, L., Karger-Kocsis, J., Kmetty, A., Drakopoulos, S.X., Production and characterization of microfibrillated cellulose-reinforced thermoplastic starch composites. *J. Appl. Polym. Sci.*, 133, 42397–42398, 2016.
10. Xia, C.L., Wang, L., Dong, Y.M., Zhang, S.F., Shi, S.Q., Cai, L.P., *et al.*, Soy protein isolate-based films cross-linked by epoxidized soybean oil. *RSC Adv.*, 5, 82765–82771, 2015.
11. El Achaby, M., El Miri, N., Snik, A., Zahouily, M., Abdelouahdi, K., Fihri, A., *et al.*, Mechanically strong nanocomposite films based on highly filled carboxymethyl cellulose with graphene oxide. *J. Appl. Polym. Sci.*, 133, 42356–42358, 2016.
12. Benbettaïeb, N., Karbowiak, T., Brachais, C.-H., Debeaufort, F., Coupling tyrosol, quercetin or ferulic acid and electron beam irradiation to cross-link chitosan–gelatin films: A structure–function approach. *Eur. Polym. J.*, 67, 113–127, 2015.
13. Li, L., Xiong, Y., Yu, S., Liu, S., Liu, F., Xie, C., Facile preparation for robust and freestanding silk fibroin films in a 1-butyl-3-methyl imidazolium acetate ionic liquid system. *J. Appl. Polym. Sci.*, 132, 42822–42827, 2015.
14. Jeoung, E., Duncan, B., Wang, L.S., Saha, K., Subramani, C., Wang, P., *et al.*, Fabrication of Robust Protein Films Using Nanoimprint Lithography. *Adv. Mat.*, 27, 6251–6255, 2015.
15. Petersen, K., Nielsen, P.V., Bertelsen, G., Lawther, M., Olsen, M.B., Nilsson, N.H., *et al.*, Potential of biobased materials for food packaging. *Trends Food Sci. Technol.*, 10, 52–68, 1999.
16. Ray, S.S., Bousmina, M., Biodegradable polymers and their layered silicate nanocomposites: in greening the 21st century materials world. *Prog. Mater. Sci.*, 50, 962–1079, 2005.
17. Zhang, H., Mittal, G., Biodegradable protein-based films from plant resources: a review. *Environ. Prog. Sustain. Energy*, 29, 203–220, 2010.
18. Gupta, P., Nayak, K.K., Characteristics of protein-based biopolymer and its application. *Polym. Eng. Sci.*, 55, 485–498, 2015.
19. Cheftel, J.C., Cuq, J.L., Lorient, D., Amino acids, peptides, and proteins. *Food Chem.*, 2, 245–369, 1985.
20. Silva, N.H.C.S., Vilela, C., Marrucho, I.M., Freire, C.S.R., Neto, C.P., Silvestre, A.J.D., Protein-based materials: from sources to innovative sustainable materials for biomedical applications. *J. Mater. Chem. B*, 2, 3715–3740, 2014.
21. Kunte, L.A., Gennadios, A., Cuppett, S.L., Hanna, M.A., Weller, C.L., Cast films from soy protein isolates and fractions 1. *Cereal Chem.*, 74, 115–118, 1997.

22. Cho, S.Y., Rhee, C., Mechanical properties and water vapor permeability of edible films made from fractionated soy proteins with ultrafiltration. *LWT-Food Sci. Technol.*, 37, 833–839, 2004.
23. Gennadios, A., Brandenburg, A.H., Weller, C.L., Testin, R.F., Effect of pH on properties of wheat gluten and soy protein isolate films. *J. Agric. Food. Chem.*, 41, 1835–1839, 1993.
24. Li, H., Li, C., Gao, Q., Zhang, S., Li, J., Properties of soybean-flour-based adhesives enhanced by attapulgite and glycerol polyglycidyl ether. *Ind. Crops Prod.*, 59, 35–40, 2014.
25. Li, C., Li, H., Zhang, S., Li, J., Preparation of reinforced soy protein adhesive using silane coupling agent as an enhancer. *Bioresources*, 9, 5448–5460, 2014.
26. Klüver, E., Meyer, M., Thermoplastic processing, rheology, and extrudate properties of wheat, soy, and pea proteins. *Polym. Eng. Sci.*, 55, 1912–1919, 2015.
27. Koshy, R.R., Mary, S.K., Thomas, S., Pothan, L.A., Environment friendly green composites based on soy protein isolate – a review. *Food Hydrocolloid.*, 50, 174–192, 2015.
28. Hammann, F., Schmid, M., Determination and quantification of molecular interactions in protein films: a review. *Materials*, 7, 7975–7996, 2014.
29. Kinsella, J.E., Functional properties of soy proteins. *J. Am. Oil Chem. Soc.*, 56, 242–258, 1979.
30. Ciannamea, E.M., Stefani, P.M., Ruseckaite, R.A., Physical and mechanical properties of compression molded and solution casting soybean protein concentrate based films. *Food Hydrocolloid.*, 38, 193–204, 2014.
31. Mauri, A.N., Añón, M.C., Effect of solution pH on solubility and some structural properties of soybean protein isolate films. *J. Sci. Food Agric.*, 86, 1064–72, 2006.
32. Liu, Y., Li, K., Development and characterization of adhesives from soy protein for bonding wood. *Int. J. Adhes. Adhes.*, 27, 59–67, 2007.
33. Zhong, Z., Sun, X.S., Properties of soy protein isolate/polycaprolactone blends compatibilized by methylene diphenyl diisocyanate. *Polymer*, 42, 6961–6969, 2001.
34. Song, X., Zhou, C., Fu, F., Chen, Z., Wu, Q., Effect of high-pressure homogenization on particle size and film properties of soy protein isolate. *Ind. Crops Prod.*, 43, 538–544, 2013.
35. Chen, L., Remondetto, G., Rouabhia, M., Subirade, M., Kinetics of the breakdown of cross-linked soy protein films for drug delivery. *Biomaterials*, 29, 3750–3756, 2008.
36. Vaz, C.M., van Doeveren, P.F.N.M., Yilmaz, G., De Graaf, L.A., Reis, R.L., Cunha, A.M., Processing and characterization of biodegradable soy plastics: effects of crosslinking with glyoxal and thermal treatment. *J. Appl. Polym. Sci.*, 97, 604–610, 2005.
37. Bigi, A., Cojazzi, G., Panzavolta, S., Rubini, K., Roveri, N., Mechanical and thermal properties of gelatin films at different degrees of glutaraldehyde crosslinking. *Biomaterials*, 22, 763–768, 2001.

38. Marquié, C., Chemical reactions in cottonseed protein cross-linking by formaldehyde, glutaraldehyde, and glyoxal for the formation of protein films with enhanced mechanical properties. *J. Agric. Food. Chem.*, 49, 4676–4681, 2001.
39. Strauss, G., Gibson, S.M., Plant phenolics as cross-linkers of gelatin gels and gelatin-based coacervates for use as food ingredients. *Food Hydrocolloid.*, 18, 81–89, 2004.
40. Reddy, D.J.P., Rajulu, A.V., Arumugam, V., Naresh, M.D., Muthukrishnan, M., Effects of resorcinol on the mechanical properties of soy protein isolate films. *J. Plast. Film Sheet.*, 25, 221–233, 2009.
41. Ma, L., Yang, Y., Yao, J., Shao, Z., Chen, X., Robust soy protein films obtained by slight chemical modification of polypeptide chains. *Polym. Chem.*, 4, 5425, 2013.
42. Shih, F.F., Interaction of soy isolate with polysaccharide and its effect on film properties. *J. Am. Oil Chem. Soc.*, 71, 1281–1285, 1994.
43. Su, J.F., Huang, Z., Liu, K., Fu, L.L., Liu, H.R., Mechanical properties, biodegradation and water vapor permeability of blend films of soy protein isolate and poly (vinyl alcohol) compatibilized by glycerol. *Polym. Bull.*, 58, 913–921, 2007.
44. John, J., Tang, J., Bhattacharya, M., Processing of biodegradable blends of wheat gluten and modified polycaprolactone. *Polymer*, 39, 2883–2895, 1998.
45. Zhang, J.W., Jiang, L., Zhu, L.Y., Jane, J.l., Mungara, P., Morphology and properties of soy protein and polylactide blends. *Biomacromolecules*, 7, 1551–1561, 2006.
46. Chen, Y., Zhang, L., Du, L., Structure and properties of composites compression-molded from polyurethane prepolymer and various soy products. *Ind. Eng. Chem. Res.*, 42, 6786–6794, 2003.
47. Zheng, H., Tan, Z., Ran Zhan, Y., Huang, J., Morphology and properties of soy protein plastics modified with chitin. *J. Appl. Polym. Sci.*, 90, 3676–3682, 2003.
48. Park, S.K., Hettiarachchy, N.S., Were, L., Degradation behavior of soy protein-wheat gluten films in simulated soil conditions. *J. Agric. Food. Chem.*, 48, 3027–3031, 2000.
49. Jongjareonrak, A., Benjakul, S., Visessanguan, W., Prodpran, T., Tanaka, M., Characterization of edible films from skin gelatin of brownstripe red snapper and bigeye snapper. *Food Hydrocolloid.*, 20, 492–501, 2006.
50. Bertini, F., Canetti, M., Patrucco, A., Zoccola, M., Wool keratin-polypropylene composites: properties and thermal degradation. *Polym. Degrad. Stab.*, 98, 980–987, 2013.
51. Audic, J.L., Chaufer, B., Daufin, G., Non-food applications of milk components and dairy co-products: a review. *Le Lait*, 83, 417–438, 2003.
52. Khwaldia, K., Arab-Tehrany, E., Desobry, S., Biopolymer coatings on paper packaging materials. *Compr. Rev. Food Sci. Food Saf.*, 9, 82–91, 2010.

53. Vepari, C., Kaplan, D.L., Silk as a biomaterial. *Prog. Polym. Sci.*, 32, 991–1007, 2007.
54. Cuq, B., Gontard, N., Guilbert, S., Proteins as agricultural polymers for packaging production. *Cereal Chem.*, 75, 1–9, 1998.
55. Panyam, D., Kilara, A., Enhancing the functionality of food proteins by enzymatic modification. *Trends Food Sci. Technol.*, 7, 120–125, 1996.
56. Guerrero, P., Stefani, P.M., Ruseckaite, R.A., de la Caba, K., Functional properties of films based on soy protein isolate and gelatin processed by compression molding. *J. Food Eng.*, 105, 65–72, 2011.
57. Guerrero, P., Garrido, T., Leceta, I., de la Caba, K., Films based on proteins and polysaccharides, Preparation and physical–chemical characterization. *Eur. Polym. J.*, 49, 3713–3721, 2013.
58. Cho, S.Y., Park, J.W., Batt, H.P., Thomas, R.L., Edible films made from membrane processed soy protein concentrates. *LWT-Food Sci. Technol.*, 40, 418–423, 2007.
59. Wang, Z., Sun, X.X., Lian, Z.X., Wang, X.X., Zhou, J., Ma, Z.S., The effects of ultrasonic/microwave assisted treatment on the properties of soy protein isolate/microcrystalline wheat-bran cellulose film. *J. Food Eng.*, 114, 183–191, 2013.
60. Ullsten, N.H., Gällstedt, M., Johansson, E., Gräslund, A., Hedenqvist, M.S., Enlarged processing window of plasticized wheat gluten using salicylic acid. *Biomacromolecules*, 7, 771–776, 2006.
61. Denavi, G., Tapia-Blácido, D.R., Añón, M.C., Sobral, P.J.A., Mauri, A.N., Menegalli, F.C., Effects of drying conditions on some physical properties of soy protein films. *J. Food Eng.*, 90, 341–349, 2009.
62. Shi, W., Dumont, M.J., Review: bio-based films from zein, keratin, pea, and rapeseed protein feedstocks. *J. Mater. Sci.*, 49, 1915–1930, 2014.
63. Fichtali, J., Van de Voort, F.R., Fundamental and practical aspects of twin screw extrusion. *Cereal Foods World*, 34, 921–929, 1989.
64. Zárate-Ramírez, L.S., Martínez, I., Romero, A., Partal, P., Guerrero, A., Wheat gluten-based materials plasticised with glycerol and water by thermoplastic mixing and thermomoulding. *J. Sci. Food Agric.*, 91, 625–633, 2011.
65. González-Gutiérrez, J., Partal, P., García-Morales, M., Gallegos, C., Effect of processing on the viscoelastic, tensile and optical properties of albumen/starch-based bioplastics. *Carbohydr. Polym.*, 84, 308–315, 2011.
66. Vieira, M.G.A., da Silva, M.A., dos Santos, L.O., Beppu, M.M., Natural-based plasticizers and biopolymer films: A review. *Eur. Polym. J.*, 47, 254–263, 2011.
67. Hauck, B.W., Huber, G.R., Single screw vs twin screw extrusion. *Cereal Foods World (USA)*, 1989.
68. Zhang, J., Mungara, P., Jane, J.-l., Mechanical and thermal properties of extruded soy protein sheets. *Polymer*, 42, 2569–2578, 2001.
69. Hernandez-Izquierdo, V.M., Krochta, J.M., Thermoplastic processing of proteins for film formation-a review. *J. Food Sci.*, 73, 30–39, 2008.

70. Guerrero, P., de la Caba, K., Thermal and mechanical properties of soy protein films processed at different pH by compression. *J. Food Eng.*, 100, 261–269, 2010.
71. Fernandes, E.M., Pires, R.A., Mano, J.F., Reis, R.L., Bionanocomposites from lignocellulosic resources: Properties, applications and future trends for their use in the biomedical field. *Prog. Polym. Sci.*, 38, 1415–1441, 2013.
72. Kumar, R., Zhang, L., Soy protein films with the hydrophobic surface created through non-covalent interactions. *Ind. Crops Prod.*, 29, 485–494, 2009.
73. Guerrero, P., Retegi, A., Gabilondo, N., de la Caba, K., Mechanical and thermal properties of soy protein films processed by casting and compression. *J. Food Eng.*, 100, 145–151, 2010.
74. Huang, X., Netravali, A., Characterization of flax fiber reinforced soy protein resin based green composites modified with nano-clay particles. *Compos. Sci. Technol.*, 67, 2005–2014, 2007.
75. Zhong, Z., Sun, X.S., Wang, D., Isoelectric pH of polyamide–epichlorohydrin modified soy protein improved water resistance and adhesion properties. *J. Appl. Polym. Sci.*, 103, 2261–2270, 2007.
76. Fombuena, V., Sánchez-Nácher L., Samper M.D., Juárez D., Balart R. Study of the properties of thermoset materials derived from epoxidized soybean oil and protein fillers. *J. Am. Oil Chem. Soc.*, 90, 449–457, 2012.
77. González, A., Strumia, M.C., Alvarez Igarzabal, C.I., Cross-linked soy protein as material for biodegradable films: Synthesis, characterization and biodegradation. *J. Food Eng.*, 106, 331–338, 2011.
78. Mauri, A.N., Anon, M.C., Mechanical and physical properties of soy protein films with pH-modified microstructures. *Food Sci. Technol. Int.*, 14, 119–125, 2008.
79. Cho, S.Y., Rhee, C., Sorption characteristics of soy protein films and their relation to mechanical properties. *LWT-Food Sci. Technol.*, 35, 151–157, 2002.
80. Ogale, A.A., Cunningham, P., Dawson, P.L., Acton, J.C., Viscoelastic, thermal, and microstructural characterization of soy protein isolate films. *J. Food Sci.*, 65, 672–679, 2000.
81. Gao, C., Stading, M., Wellner, N., Parker, M.L., Noel, T.R., Mills, E.N.C., *et al.*, Plasticization of a protein-based film by glycerol: a spectroscopic, mechanical, and thermal study. *J. Agric. Food. Chem.*, 54, 4611–4616, 2006.
82. Cuq, B., Gontard, N., Guilbert, S., Thermal properties of fish myofibrillar protein-based films as affected by moisture content. *Polymer*, 38, 2399–2405, 1997.
83. Cuq, B., Gontard, N., Guilbert, S., Thermoplastic properties of fish myofibrillar proteins: application to biopackaging fabrication. *Polymer*, 38, 4071–4078, 1997.
84. Mo, X., Sun, X., Plasticization of soy protein polymer by polyol-based plasticizers. *J. Am. Oil Chem. Soc.*, 79, 197–202, 2002.
85. Guo, G., Zhang, C., Du, Z., Zou, W., Tian, H., Xiang, A., *et al.*, Structure and property of biodegradable soy protein isolate/PBAT blends. *Ind. Crops Prod.*, 74, 731–736, 2015.

86. Kokoszka, S., Debeaufort, F., Hambleton, A., Lenart, A., Voilley, A., Protein and glycerol contents affect physico-chemical properties of soy protein isolate-based edible films. *Innov. Food Sci. Emerg. Technol.*, 11, 503–510, 2010.
87. Di Gioia, L., Guilbert, S., Corn protein-based thermoplastic resins: effect of some polar and amphiphilic plasticizers. *J. Agric. Food. Chem.*, 47, 1254–1261, 1999.
88. Unakar, N.J., Tsui, J.Y., Harding, C.V., Scanning electron microscopy. *Ophthalmic Res.*, 13, 20–35, 1981.
89. Li, Y., Chen, F., Zhang, L., Yao, Y., Effect of surface changes of soy protein materials on water resistance. *Mater. Lett.*, 149, 120–122, 2015.
90. Murray, B.S., Rheological properties of protein films. *Curr. Opin. Colloid Interf. Sci.*, 16, 27–35, 2011.
91. Wang, J.M., Xia, N., Yang, X.Q., Yin, S.W., Qi, J.R., He, X.T., *et al.*, Adsorption and dilatational rheology of heat-treated soy protein at the oil-water interface: relationship to structural properties. *J. Agric. Food Chem.*, 60, 3302–3310, 2012.
92. Sun, L., Chen, W., Liu, Y., Li, J., Yu, H., Soy protein isolate/cellulose nanofiber complex gels as fat substitutes: rheological and textural properties and extent of cream imitation. *Cellulose*, 22, 2619–2627, 2015.
93. Blanchet, C.E., Svergun, D.I., Small-angle X-ray scattering on biological macromolecules and nanocomposites in solution. *Ann. Rev. Phys. Chem.*, 64, 37–54, 2013.
94. Pérez, J., Nishino, Y., Advances in X-ray scattering: from solution SAXS to achievements with coherent beams. *Curr. Opin. Struct. Biol.*, 22, 670–678, 2012.
95. Graewert, M.A., Svergun, D.I., Impact and progress in small and wide angle X-ray scattering (SAXS and WAXS). *Curr. Opin. Struct. Biol.*, 23, 748–754, 2013.
96. O. Glatter, O. Kratky, Eds., *Small angle X-ray scattering*, Academic Press: London, 1982.
97. Lai, H.M., Geil, P.H., Padua, G.W., X-ray diffraction characterization of the structure of zein–Oleic acid films. *J. Appl. Polym. Sci.*, 71, 1267–1281, 1999.
98. Tummala, P., Liu, W., Drzal, L.T., Mohanty, A.K., Misra, M., Influence of plasticizers on thermal and mechanical properties and morphology of soy-based bioplastics. *Ind. Eng. Chem. Res.*, 45, 7491–7496, 2006.
99. Liu, W., Misra, M., Askeland, P., Drzal, L.T., Mohanty, A.K., 'Green' composites from soy based plastic and pineapple leaf fiber: fabrication and properties evaluation. *Polymer*, 46, 2710–2721, 2005.
100. Huang, J., Zhang, L.N., Chen, F.G., Effects of lignin as a filler on properties of soy protein plastics. 1. Lignosulfonate. *J. Appl. Polym. Sci.*, 88, 3284–3290, 2003.
101. Wei, M., Fan, L.H., Huang, J., Chen, Y., Role of star-like hydroxylpropyl lignin in soy-protein plastics. *Macromol. Mater. Eng.*, 291, 524–530, 2006.

102. Lu, Y.S., Weng, L.H., Zhang, L.N., Morphology and properties of soy protein isolate thermoplastics reinforced with chitin whiskers. *Biomacromolecules*, 5, 1046–1051, 2004.
103. Wang, Y.X., Cao, X.D., Zhang, L.N., Effects of cellulose whiskers on properties of soy protein thermoplastics. *Macromol. Biosci.*, 6, 524–531, 2006.
104. Lu, Y., Chen, W., Sub-nanometre sized metal clusters: from synthetic challenges to the unique property discoveries. *Chem. Soc. Rev.*, 41, 3594–3623, 2012.
105. Takemori, R., Ito, G., Tanaka, M., Kawakami, H., Ultra-high proton conduction in electrospun sulfonated polyimide nanofibers. *RSC Adv.*, 4, 20005–20009, 2014.
106. Shang, L., Dong, S., Nienhaus, G.U., Ultra-small fluorescent metal nanoclusters: synthesis and biological applications. *Nano Today*, 6, 401–418, 2011.
107. Xu, H., Suslick, K.S., Water-soluble fluorescent silver nanoclusters. *Adv. Mater.*, 22, 1078–1082, 2010.
108. Ai, F.J., Zheng, H., Wei, M., Huang, J., Soy protein plastics reinforced and toughened by SiO_2 nanoparticles. *J. Appl. Polym. Sci.*, 105, 1597–1604, 2007.
109. Li, K., Chen, H., Li, Y., Li, J., He, J., Endogenous Cu and Zn nanocluster-regulated soy protein isolate films: excellent hydrophobicity and flexibility. *RSC Adv.*, 5, 66543–66548, 2015.
110. Luo, L., Gan, L., Liu, Y., Tian, W., Tong, Z., Wang, X., *et al.*, Construction of nerve guide conduits from cellulose/soy protein composite membranes combined with Schwann cells and pyrroloquinoline quinone for the repair of peripheral nerve defect. *Biochem. Biophys. Res. Commun.*, 457, 507–513, 2015.
111. Luo, L., Gong, W., Zhou, Y., Yang, L., Li, D., Huselstein, C., *et al.*, Cellulose/soy protein isolate composite membranes: evaluations of *in vitro* cytocompatibility with Schwann cells and *in vivo* toxicity to animals. *Biomed. Mater. Eng.*, 25, 57–64, 2015.
112. Ozcalik, O., Tihminlioglu, F., Barrier properties of corn zein nanocomposite coated polypropylene films for food packaging applications. *J. Food Eng.*, 114, 505–513, 2013.
113. Wang, H., Jiang, Q., Jiang, S., Jiang, S., Fabrication and characterization of PVA/SPI/SiO_2 hybrid fibres via electrospinning technique. *Polym. Polym. Compos.*, 20, 621–627, 2012.
114. Wang, S.Y., Zhu, B.B., Li, D.Z., Fu, X.Z., Shi, L., Preparation and characterization of TIO2/SPI composite film. *Mater. Lett.*, 83, 42–45, 2012.
115. Wang, Z., Zhang, N., Wang, H.-y., Sui, S.-y., Sun, X.-x., Ma, Z.-s., The effects of ultrasonic/microwave assisted treatment on the properties of soy protein isolate/titanium dioxide films. *LWT - Food Sci. Technol.*, 57, 548–555, 2014.

116. Voon, H.C., Bhat, R., Easa, A.M., Liong, M.T., Karim, A.A., Effect of addition of halloysite nanoclay and SiO_2 nanoparticles on barrier and mechanical properties of Bovine gelatin films. *Food Bioprocess Tech.*, 5, 1766–1774, 2010.
117. Lubasova, D., Mullerova, J., Netravali, A.N., Water-resistant plant protein-based nanofiber membranes. *J. Appl. Polym. Sci.*, 132, 41852–41859, 2015.
118. Jensen, A., Lim, L.T., Barbut, S., Marcone, M., Development and characterization of soy protein films incorporated with cellulose fibers using a hot surface casting technique. *LWT - Food Sci. Technol.*, 60, 162–170, 2015.
119. Avena-Bustillos, R.J., Krochta, J.M., Water vapor permeability of caseinate-based edible films as affected by pH, calcium crosslinking and lipid content. *J. Food Sci.*, 58, 904–907, 1993.
120. Kester, J.J., Richardson, T., Modification of whey proteins to improve functionality. *J. Dairy Sci.*, 67, 2757–2774, 1984.
121. Liu, L., Chen, Z., Wang, B., Yang, Y., Improving wet strength of soy protein films using oxidized sucrose. *J. Appl. Polym. Sci.*, 132, 41473–41477, 2015.
122. Su, J.F., Yuan, X.Y., Huang, Z., Wang, X.Y., Lu, X.Z., Zhang, L.D., *et al.*, Physicochemical properties of soy protein isolate/carboxymethyl cellulose blend films crosslinked by Maillard reactions: color, transparency and heat-sealing ability. *Mater. Sci. Eng. C Mater. Biol. Appl.*, 32, 40–46, 2012.
123. Ustunol, Z., Mert, B., Water solubility, mechanical, barrier, and thermal properties of cross-linked whey protein isolate-based films. *J. Food Sci.*, 69, 129–133, 2004.
124. Cherian, G., Gennadios, A., Weller, C.L., Chinachoti, P., Thermomechanical behavior of wheat gluten films: effect of sucrose, glycerin, and sorbitol. *Cereal Chem.*, 72, 1–6, 1995.
125. Park, S.K., Bae, D.H., Rhee, K.C., Soy protein biopolymers cross-linked with glutaraldehyde. *J. Am. Oil Chem. Soc.*, 77, 879–884, 2000.
126. Yao, C., Li, X., Song, T., Electrospinning and crosslinking of zein nanofiber mats. *J. Appl. Polym. Sci.*, 103, 380–385, 2007.
127. Dieteroch, D., Grigat, E., Hahn, W., *Chemical and physical-chemical principles of polyurethane chemistry*. Polyurethane Handbook Hanser, New York, 1985.
128. Fang, K., Wang, B., Sheng, K., Sun, X.S., Properties and morphology of poly(lactic acid)/soy protein isolate blends. *J. Appl. Polym. Sci.*, 114, 754–759, 2009.
129. Lei, H., Du, G., Wu, Z., Xi, X., Dong, Z., Cross-linked soy-based wood adhesives for plywood. *Int. J. Adhes. Adhes.*, 50, 199–203, 2014.
130. Gao, Z.H., Zhang, Y.H., Fang, B., Zhang, L.P., Shi, J.Y., The effects of thermal-acid treatment and crosslinking on the water resistance of soybean protein. *Ind. Crops Prod.*, 74, 122–131, 2015.
131. McHugh, T.H., Krochta, J.M., Sorbitol-vs glycerol-plasticized whey protein edible films: integrated oxygen permeability and tensile property evaluation. *J. Agric. Food. Chem.*, 42, 841–845, 1994.
132. Valencia-Chamorro, S.A., Palou, L., Del Río, M.A., Pérez-Gago, M.B., Antimicrobial edible films and coatings for fresh and minimally processed fruits and vegetables: a review. *Crit. Rev. Food Sci. Nutr.*, 51, 872–900, 2011.

133. Falguera, V., Quintero, J.P., Jiménez, A., Muñoz, J.A., Ibarz, A., Edible films and coatings: structures, active functions and trends in their use. *Trends Food Sci. Technol.*, 22, 292–303, 2011.
134. González, A., Igarzabal, C.I.A., Nanocrystal-reinforced soy protein films and their application as active packaging. *Food Hydrocolloid.*, 43, 777–784, 2015.
135. Donhowe, I.G., Fennema, O., Edible films and coatings: characteristics, formation, definitions, and testing methods, in *Edible coatings and films to improve food quality*, 1–24, CRC Press, Boca Raton, FL 1994.
136. Guerrero, P., O'Sullivan, M.G., Kerry, J.P., de la Caba, K. Application of soy protein coatings and their effect on the quality and shelf-life stability of beef patties. *RSC Adv.*, 5, 8182–8189, 2015.
137. Usui, M., Tamura, H., Nakamura, K., Ogawa, T., Muroshita, M., Azakami, H., *et al.*, Enhanced bactericidal action and masking of allergen structure of soy protein by attachment of chitosan through Maillard-type protein-polysaccharide conjugation. *Food/Nahrung*, 48, 69–72, 2004.
138. Appendini, P., Hotchkiss, J.H., Review of antimicrobial food packaging. *Innov. Food Sci. Emerg. Technol.*, 3, 113–126, 2002.
139. Pan, Y., Huang, X., Shi, X., Zhan, Y., Fan, G., Pan, S., *et al.*, Antimicrobial application of nanofibrous mats self-assembled with quaternized chitosan and soy protein isolate. *Carbohydr. Polym.*, 133, 229–235, 2015.
140. MaHam, A., Tang, Z., Wu, H., Wang, J., Lin, Y., Protein-based nanomedicine platforms for drug delivery. *Small*, 5, 1706–1721, 2009.
141. Wu, H., Engelhard, M.H., Wang, J., Fisher, D.R., Lin, Y., Synthesis of lutetium phosphate–apoferritin core–shell nanoparticles for potential applications in radioimmunoimaging and radioimmunotherapy of cancers. *J. Mater. Chem.*, 18, 1779–1783, 2008.
142. Suci, P.A., Berglund, D.L., Liepold, L., Brumfield, S., Pitts, B., Davison, W., *et al.*, High-density targeting of a viral multifunctional nanoplatform to a pathogenic, biofilm-forming bacterium. *Chem. Biol.*, 14, 387–398, 2007.
143. Flenniken, M.L., Liepold, L.O., Crowley, B.E., Willits, D.A., Young, M.J., Douglas, T., Selective attachment and release of a chemotherapeutic agent from the interior of a protein cage architecture. *Chem. Commun.*, 447–449, 2005.
144. Furgeson, D.Y., Dreher, M.R., Chilkoti, A., Structural optimization of a "smart" doxorubicin–polypeptide conjugate for thermally targeted delivery to solid tumors. *J. Control. Rel.*, 110, 362–369, 2006.
145. Liu, X., Sun, Q., Wang, H., Zhang, L., Wang, J.-Y., Microspheres of corn protein, zein, for an ivermectin drug delivery system. *Biomaterials*, 26, 109–115, 2005.
146. Swatschek, D., Schatton, W., Müller, W.E.G., Kreuter, J., Microparticles derived from marine sponge collagen (SCMPs): preparation, characterization and suitability for dermal delivery of all-trans retinol. *Eur. J. Pharma. Biopharma.*, 54, 125–133, 2002.
147. Muvaffak, A., Gurhan, I., Hasirci, N., Prolonged cytotoxic effect of colchicine released from biodegradable microspheres. *J. Biomed. Mater. Res. Part B: Appl. Biomater.*, 71, 295–304, 2004.

148. Peles, Z., Binderman, I., Berdicevsky, I., Zilberman, M., Soy protein films for wound-healing applications: antibiotic release, bacterial inhibition and cellular response. *J. Tissue Eng. Regen. Med.*, 7, 401–412, 2013.
149. Fink, J.K., *The chemistry of bio-based polymers*, John Wiley & Sons, 2014.
150. Kasirajan, S., Ngouajio, M., Polyethylene and biodegradable mulches for agricultural applications: a review. *Agron. Sustain. Dev.*, 32, 501–529, 2012.
151. Brodhagen, M., Peyron, M., Miles, C., Inglis, D.A., Biodegradable plastic agricultural mulches and key features of microbial degradation. *Appl. Microbiol. Biotechnol.*, 99, 1039–1056, 2015.
152. Yang, N., Sun, Z.-X., Feng, L.-S., Zheng, M.-Z., Chi, D.-C., Meng, W.-Z., *et al.*, Plastic film mulching for water-efficient agricultural applications and degradable films materials development research. *Mater. Manuf. Process.*, 30, 143–154, 2015.
153. Rhim, J.W., Lee, J.H., Effect of $CaCl_2$ treatment on mechanical and moisture barrier properties of sodium alginate and soy protein-based films. *Food Sci. Biotechnol.*, 13, 728–732, 2004.

9

Use of Soy Protein-Based Carriers for Encapsulating Bioactive Ingredients

Zhen-Xing Tang* **and Jie-Yu Liang**

Hangzhou Tianlong Group Co. Ltd, Hangzhou, China

Abstract
The applications of food-grade encapsulation carriers for protection and controlled release of bioactive food ingredients have been used in functional foods fields. Soy proteins are renewable, biodegradable, and abundant raw materials with interesting functional and/or physicochemical properties, which can be made into various encapsulation carrier forms, such as micro- and nanoparticles, fibers, films, hydrogels, etc. In most cases, encapsulation methods, such as spray-drying, coacervation, extrusion, emulsion, etc., are mainly used for the preparation of soy protein-based encapsulation carriers. In this chapter, the recent progress in the preparation of soy protein-based carriers for bioactive ingredients encapsulation was reviewed.

Keywords: Encapsulation, soy protein, modification, bioactive ingredients

9.1 Introduction

Food ingredients perform many roles when they are formulated and processed into food products. They not only contribute to flavour and texture but also contribute to some food ingredients (vitamins, minerals, folic acid, phytochemicals and functional lipids, probiotics, and proteins) and have a physiological role. These ingredients, also known as bioactive ingredients, have been used in the production of functional foods [1–3]. However, many of these bioactive ingredients cannot be simply introduced into foods

**Corresponding author*: tangzhenxing@126.com

Visakh P. M and Olga Nazarenko (eds.) Soy Protein-Based Blends, Composites and Nanocomposites, (231–250) 2017 © Scrivener Publishing LLC

in pure form. For instance, they have poor solubility in aqueous or lipid phase and are sensitive to food processing and storage conditions (temperature, light, and oxygen) as well as digestive conditions in the human gastrointestinal tract [3–6]. It will compromise the overall functionality of foods, particularly hindering food sensory properties and decreasing the bioavailability of bioactive ingredients [1, 7–9]. Therefore, in order to incorporate bioactive ingredients into the foods, it must be protected during the development of functional foods and also during gastrointestinal transit until they reach the desired sites in the body after ingestion [10, 11].

The use of encapsulation technology for protecting bioactive ingredients has been received increasing attention [7, 10, 12]. Encapsulation technology has many applications in the medical, pharmaceutical, cosmetics, chemical, agricultural, and food industries [1, 13, 14]. Encapsulation is a method that can package small particles of bioactive ingredients, within the matrix or wall material, and thus forming small capsules. Bioactive ingredients in the capsule are isolated from the surrounding environment and can be released in response to a trigger such as shear, pH, or enzyme action, thus enabling their controlled delivery to a desired site.

The encapsulation materials used in food applications must be selected from natural biomaterials or compounds with granted generally recognized as safe (GRAS) status. Food biopolymers, such as proteins, carbohydrates, are the most popular wall materials used for encapsulating bioactive ingredients [2,3,6,7]. Among these food-grade encapsulation materials, food proteins are biopolymers that have considerable functional properties, such as emulsification, gelation, foaming, etc [2–4,15,16]. Currently, proteins extracted from animal-derived products (whey proteins, gelatin, casein) and from vegetables (soy proteins, pea proteins, cereal proteins) have been widely used for encapsulation of bioactive ingredients. Compared with animal-derived proteins, the use of vegetable proteins as encapsulation materials shows the present "green" conception in food industry. In food applications, vegetable proteins are known to be less allergenic compared to animal-derived proteins [17,18]. Furthermore, vegetable proteins not only provide micronutrients to humans but also offer some health benefits that have not been found in animal-derived proteins [3,19,20]. Among vegetable proteins, soy proteins are most prevalent, which have been used as encapsulation materials alone or in combination with polysaccharide depending on the applications. Soy proteins are obtained from soy bean that contain a fraction (35–40%) of proteins, mainly glycinin and conglycin (50–90% of total proteins) [21]. Molecular weight of glycinin fraction (11S globulin) is about 350 kDa while conglycin fraction (7S globulin) is about 70 kDa. Soy proteins have interesting physicochemical and functional properties, particularly including gel-forming, emulsifying, and surfactant properties [2,3,22]. Soy proteins' characteristics and their solubility are strongly dependent on pH, heat treatment, and the presence and concentration of salts or other ingredients (oil, carbohydrate, surfactant).

Development of soy protein-based encapsulation agents not only provides opportunities to offer novel functional foods to consumers, particularly for vegan diets, but also meets the present sustainable and green trends in food fields. For these reasons, many efforts have been made to explore the possibilities of utilizing soy proteins as encapsulation carriers for bioactive ingredients in past few years [2,3,23]. Hence, in this chapter, the recent progress on soy protein-based encapsulation carriers for bioactive ingredients was reviewed.

9.2 Encapsulation Methods

Many methods can be used to produce encapsulated bioactive ingredients. Among these encapsulation methods, the main methods for producing soy protein-based carriers are spray drying, extrusion, emulsion and coacervation, etc [2, 3].

9.2.1 Spray Drying

Spray drying has been used for the production of many encapsulated bioactive ingredients—vitamins, enzymes, minerals, polyunsaturated oils, flavors, and probiotics [24]. Spray drying is a continuous process, which involves dissolving the core in a dispersion of wall material and atomizing into heated air in drying chamber to remove water, and thus the encapsulated powder is formed. Advantages of this method are simple, relatively inexpensive, and rapid. Main disadvantage of this technology is the possibility of degradation of bioactive ingredients at high drying temperature. The important factor for spray drying is a high solubility of wall material in water (or other chosen solvent) and a low viscosity at high solid content [2].

9.2.2 Extrusion

The basic process for extrusion is to force a hot biopolymer mass containing the dispersed bioactive ingredients through an orifice into a hardening bath. This process has been extensively used to encapsulate many bioactive ingredients [24–25]. Syringe extrusion is the most used method in the research for the formation of alginate beads. Alginate solution containing bioactive core is extruded as droplets into a calcium chloride solution, and thus alginate based beads are formed [24].

9.2.3 Complex Coacervation

The process for complex coacervate is that the polyelectrolyte mixtures from a solution is separated and deposited onto the agglomerated colloidal particles (i.e., wall material). Complex coacervates of oppositely charged biopolymers are the most used method in the food industry for encapsulating bioactive ingredients [2]. Many factors including the biopolymer type (molar mass, flexibility, and charge) concentration and the ratio of the biopolymers, ionic strength, and pH of the solution affect the formation of complex coacervates [25–28]. Although electrostatic interactions are considered to drive the interaction between biopolymers of opposite charge, hydrophobic interactions and hydrogen bonding can also contribute significantly to the formation of complex coacervates [25,28]. One limitation of this method is that complex coacervates are sensitivity to pH and ionic strength. To increase the robustness of coacervates, the formed coacervates need to be cross-linked after coacervates are formed. This work plays an important role in keeping coacervates stability. Glutaraldehyde is the most commonly used cross-linking agent [29]. However, glutaraldehyde is a relatively toxic solvent, which limits its use in food applications [30, 31].

9.2.4 Emulsion-Based Systems

Emulsion-based systems have been used for encapsulating bioactive food ingredients (e.g., omega-3 oils, carotenes, tocopherols). Emulsion-based encapsulation systems with different functional properties can be designed by tailoring the characteristics of the dispersed phase (e.g., size, charge, interfacial properties of droplets) as well as the microstructure of the emulsions [10, 25]. As we know, many bioactive ingredients are lipophilic. Therefore, oil-in-water (O/W) emulsion is developed to deliver these lipophilic substances. Currently, O/W emulsions are the most widely used emulsion systems for encapsulating lipophilic bioactive ingredients. The basic process for O/W emulsion is that it solubilizes the hydrophobic bioactive ingredients in a lipid phase and then homogenizes them with an aqueous phase containing food grade emulsifiers [10,11,25]. In recent years, the emulsions stabilized by food-grade particles, also called as "pickering emulsion," have been receiving a growing interest in food fields. Pickering emulsion can provide outstanding physical and chemical stability to encapsulated bioactive ingredients [32–33]. However, the application of particles derived from food proteins as the emulsifier is just at the initial stage [3].

9.3 Soy Protein-Based Encapsulation Carriers

9.3.1 Improving the Performance of Soy Proteins as Wall Materials

Soy proteins have relatively lower solubility in aqueous systems than animal-derived proteins do [34–37]. Therefore, in order to improve their solubility, and subsequently emulsifying properties and encapsulating abilities, it is highly necessary to modify soy proteins. There has been extensive work on the modification of soy proteins in the literature aimed at improving functionality [2,3]. Jiang *et al.* [38] modified the native structure of soy protein by pretreating in acid (pH 1.5–3.5) and alkaline (pH 10.0–12.0) solutions for various times (0–4 h). The authors found the pH pretreatments could increase the surface hydrophobicity of soy protein and thus resulted in a significant improvement of its emulsifying properties. Augustin *et al.* [39] studied the effect of heat treatments on mixtures of proteins (sodium caseinate, whey protein isolate, and soy protein) and carbohydrates (glucose, dried glucose syrup, oligosaccharide) for the encapsulation of fish oil. The results showed that the obtained protein–carbohydrate mixtures after treatment at high temperature before emulsification improved the oxidative stability of fish oil microcapsules. Additionally, the mixtures of protein–carbohydrate (or polysaccharides) can enhance the emulsion stability [40]. Nesterenko *et al.* [6] investigated the effects of enzymatic hydrolysis and N-acylation on the encapsulation properties of soy protein isolate (SPI) using α-tocopherol as the bioactive ingredient. O/W emulsions containing different amounts of native or modified SPI and α-tocopherol were spray dried. It was found that hydrolysis and N-acylation resulted in a decrease of emulsion droplet size and viscosity. Hydrolyzed SPI could produce the emulsions with smaller droplets and lower oil retention efficiency (39%) in the corresponding powder due to the insufficient chain length of wall material to produce a stable matrix during spray drying. On the other hand, SPI with N-acylation resulted in a significantly higher retention efficiency (>87%) compared to the efficiency obtained with native SPI (80%). This increased retention efficiency after protein acylation was observed, demonstrating that soy proteins in modified state represent a relevant better encapsulation agent for hydrophobic ingredients.

9.3.2 Encapsulation of Bioactive Ingredients Using Soy Protein-Based Carriers

Soy proteins are now one of the most widely used proteins in food processing. Soy protein-based carriers with different structures and forms, such as microparticle, nanoparticles, fiber, film hydrogel, etc., can be obtained when different processing methods are used [3,41]. Some examples are outlined in Table 9.1.

SPI microparticles can be made using spray drying, coacervation, and extrusion [2, 37, 42, 43]. Chen and Subirade prepared SPI/zein complex microspheres by extrusion method for delivering riboflavin [42–43]. The obtained particles (about 15–25μm) had spherical morphology with homogenous distribution throughout the matrix. Microspheres with SPI/zein ratios of 5:5 and 3:7 displayed near-zero-order release kinetics in simulated gastrointestinal fluids. The release of riboflavin from pure SPI

Table 9.1 Encapsulation with SPI-based wall material.

Wall materials	Bioactive ingredients	References
Mixture of proteins and polysaccarides	Fish oil	[39]
SPI/glucose syrup	Stearin, palm oil	[62]
SPI/maltrodextrin	Phospholipide	[78]
SPI	Flavors	[79]
SPI	Casein hydrolysate	[74]
SPI/gelatin	Casein hydrolysate	[73]
SPI	tocopherol	[6]
Soy glycinin	Hexadecane	[37]
Soy glycinin/sodium dodecylsufate	Hexadecane	[29]
SPI/pectin	Casein hydrolysate	[61]
SPI/pectin	propolis	[80]
SPI/gum Arabic	Orange Oil	[60]
SPI	Riboflavin	[42]

microspheres in the stomach compartment could be finished within 15 min, while the SPI/zein microspheres provided sustained release of riboflavin over 4 h. Incorporation of these microspheres into yogurt significantly delayed riboflavin release, which would increase the likelihood of gastric-sensitive nutrients reaching the intestine for absorption. The results indicated that SPI/zein microspheres showed potential as nutraceutical delivery carriers for the development of functional foods, such as yogurt enriched with vitamins [43]. Rascon *et al.* [44] investigated the performance of SPI for the encapsulation of paprika oleoresin by spray drying. The authors found that carotenoid retention in the microparticles increased as inlet air temperature was increased from 160 °C to 200 °C. The maximal stability of carotenoid oxidation in the microparticles with the highest carotenoid was found at water activity a_w of 0.743. Tang and Li [45] prepared encapsulated soy oil using SPI by spray drying. The effects of homogenization pressure (40 MPa or 80 MPa), inlet air temperature (160 °C, 180 °C, and 200°C), core: wall ratio (1:2, 1:1, 3:2, and 3:1) and total solid content (10%, 15%, and 20%) on the obtained powder properties were studied. The results showed that the core: wall ratio was the most important factor affecting retention efficiency, redispersion characteristics, and storage stability of spray-dried emulsions.

Due to the ligand binding properties, soy proteins can bind bioactive ingredients to form complexes in nanoscale through physical interactions, mainly hydrophobic interactions, hydrogen bonds, and van der Waals attraction [3]. Recently, Zhang *et al.* [46] prepared SPI nanoparticles by ethanol desolvation method. SPI exhibited average size (150nm), ζ-potential (−36mV), and high encapsulation efficiency for curcumin (97.2%). The release of encapsulated curcumin in phosphate buffer saline followed a biphasic pattern. Teng *et al.* [47] prepared folic acid (FA)-conjugated SPI nanoparticles. Compared to SPI nanoparticles without FA, FA-SPI nanoparticles showed a lower average size, a higher loading efficiency, and a faster release of curcumin in Tween 20-PBS buffer. Cellular uptake of SPI nanoparticles was increased by at most 93% in Caco-2 cells upon the conjugation with FA [47]. Zhang *et al.* [46] found that calcium-induced SPI nanoparticles (28–179nm) had uniform size distribution and spherical shape. Nanoparticle characteristics could be modulated by changing pH and calcium concentration. *In vitro* study indicated that these nanoparticles were nontoxic and mainly distributed in cytoplasm when they were absorbed into Caco-2 cells. Zhang *et al.* [48] further investigated the intestinal uptake and transport mechanisms of Ca^{2+}-induced SPI nanoparticles for vitamin B_{12} (VB_{12}). The results showed that the intestinal transport and uptake of VB_{12} were improved up to fourfold after being encapsulated into

30 nm SPI nanoparticles. Teng *et al.* [49] prepared SPI/carboxymethyl chitosan (CMCS) nanoparticles by Ca^{2+} induced method and employed as a delivery carrier for hydrophobic vitamin D_3 (VD_3). The VD_3-loaded complex nanoparticles with an average size of 162–243 nm remained stable in aqueous phase. Compared to pure SPI nanoparticles, the SPI/CMCS nanoparticles exhibited a reduced (42.3% compared to 86.1%) release of VD_3 in simulated gastric fluid and an enhanced (36.0% compared to 8.2%) release under simulated intestinal condition. In another study, Ding and Yao [50] obtained FA-loaded soy protein/soy polysaccharide complex nanogels by using high pressure homogenization and heating procedures. The nanogels were dispersible and stable after six months of storage in acidic conditions (pH 3.0–5.0). Moreover, the nanogels could protect FA from decomposition in the presence of heat, light, and oxygen at acidic conditions and release FA in the intestine. Gao *et al.* [51] found soy lipophilic protein (LP, a group of protein fractions associated with lecithin) could transform into nanoparticles in aqueous phase after an ultrasonic treatment. Subsequently, Gao *et al.* [52] incorporated conjugated-linoleic acid (CLA) into lipophilic protein nanoparticle (LPP) by ultrasonication. The CLA-loaded LPP had a mean particle diameter of 170nm and a loading capacity of 26.3%. These findings suggest LPP with a core shell structure can be used as a delivery vehicle for hydrophobic bioactive ingredients.

Soy proteins have been extensively used as an emulsifier due to its high emulsifying properties [2, 3, 53]. Many studies show that SPI-stabilized O/W emulsions could be used as encapsulating systems for protecting bioactive lipids, such as omega-3 fatty acids, and incorporating them into functional foods [3, 54, 55]. Tang and Liu [56] prepared soy protein aggregates (~100 nm) by thermal treatment and found that sodium chloride could act as emulsion stabilizer. The resulting emulsions showed extraordinary stability against coalescence and creaming. RES (trans-3, 5, 4′-trihydroxystilbene), is a natural polyphenol compound mainly found in red grapes and peanuts. Wan *et al.* [57] prepared SPI-RES complex and then used the complex as an emulsifier at the O/W interface. The resulting O/W emulsion showed an increased oxidative stability due to the high interfacial accumulation of RES. Yuan *et al.* [58] fabricated a mixed emulsion by using the soluble β-conglycinin (7S)-chitosan (CS) complex as emulsifier. The obtained 7S/CS mixed emulsion exhibited good storage stability against microorganisms at acidic pHs due to the positively charged groups of CS around oil droplet surfaces, which could interact with negatively charged cell membranes, thus resulting in the inhibition of microorganism growth. Furthermore, Wan *et al.* [3, 59] introduced a novel biosurfactant stevioside (STE) to encapsulate RES by the formation of STE self-assembled micelles.

The physical and oxidative stability of SPI-based O/W emulsion were significantly improved by the incorporation of the STE-RES complex.

Many studies have reported that a decrease in process efficiency is observed when active hydrophobic core concentration is more than 50% [3,37,60–62]. The protein concentration during the emulsification step is strongly related to the stability and size of the emulsion [37]. Due to protein surfactant properties, increasing protein concentration will result in improving the adsorption of proteins on the oil–water interface and the droplets' coalescence resistance and thus decreasing in the mean diameter of droplet. On the other hand, Lazko *et al.* [37] also demonstrated that protein concentration did not have a significant influence on the final microparticles wall thickness.

Fibers have diversified applications in food and biomedical fields. Vega-Lugo and Lim [63] prepared SPI/poly(ethylene oxide) (PEO) blend and poly(lactic acid) fibers (200nm to 2μm) containing an antimicrobial compound, allyl isothiocyanate (AITC). The release of AITC could be controlled by varying the relative humidity, and exposure to air with elevated relative humidity triggered the release of AITC. Wang *et al.* [64] successfully incorporated the anthocyanin-rich red raspberry extract into SPI-PEO composite electrospun fibers, which endowed the fibers with a high antibacterial activity against Staphylococcus epidermidis.

Hydrogels are three-dimensional networks constructed by physically or chemically cross-linked polymers, which are capable of holding a large quantity of water [3,65,66]. In this system, bioactive ingredients can be entrapped, protected from the environments. The release behavior of hydrogels can be manipulated by modifying their microstructure (67). Maltais *et al.* [68,69] compared the controlled release behavior of two types of cold-set soy protein hydrogels, filamentous, or particulate. The results showed that filamentous hydrogels exhibited a delayed release of riboflavin due to their lower porosity when compared to the particulate hydrogels. Both hydrogels provided good protection for riboflavin at least 6 h under pH1.2. The release of riboflavin was independent of time or concentration (zero-order release), when the gels were in the presence of pancreatin at pH7.5. Caillard *et al.* [70,71] investigated soy protein hydrogels cross-linked by glutaraldehyde *in vitro* as carriers for the release of ionic compounds (amaranth and methylene blue). Increasing the cross-linking degree and the concentration of salt in the gel generally led to the decrease of swelling/release rates of soy protein hydrogel. These results demonstrated the release behavior of hydrogel that can be controlled by preparation parameters of hydrogel. Controlling the interaction between protein and polysaccharide is an effective strategy for preparing pH-responsive hydrogel delivery carriers. Guo *et al.* [41,72] prepared soy glycinin /dextran sulfate (DS) transparent hydrogels cross-linked by microbial transglutaminase (MTGase). The authors found that the hydrogel deswelled in simulated gastric fluid and swelled in simulated intestinal fluid. The extent of deswelling and swelling increased when more DS was used in the hydrogel. What is more, the hydrogel showed a sustained release behavior under simulated gastrointestinal conditions.

Soy proteins are also used to mask the bitter taste of some hydrophobic hydrolysates (e.g., casein hydrolysate) [61,73,74]. In this case, the authors demonstrated the decrease in encapsulation efficiency and the increase in particle size with increasing active core concentration. Jun-xia *et al.* [60] used SPI-gum Arabic (GA) coacervates for encapsulation of sweet orange oil. Effects of the factors including pH, ionic strength, SPI/GA ratio, core material load, and addition of sucrose and maltodextrin on encapsulation efficiency were investigated. The results showed that SPI treated by ultrasonics prior to encapsulation could increase their solubility. The optimum conditions for high encapsulation efficiency were followed as: pH 4.0, 1:1 SPI/GA ratio, and 10 % core material load. It was also found that the addition of sucrose into encapsulation system increased the encapsulation yield from 65% to 67%. de Conto *et al.* [75] encapsulated omega-3 ethyl ester by complex coacervation using SPI and GA. Effect of the factors such as core: wall ratio, concentration of wall materials, and transglutaminase on the encapsulation efficiency was investigated. The highest encapsulation efficiency was observed at 2.6:1.0 wall/core ratio and 1.8:1.0 SPI/GA ratio.

Overall, many studies have showed the abilities of soy proteins as encapsulating agents. The main differences between encapsulating agents are the structure and size of the obtained carriers and consequently the release of the bioactive ingredients.

9.3.3 Our Work on Using Soy Protein-Based Carriers

The application of soy protein-based carriers in probiotics encapsulation is very limited. In our group, the mixture of alginate and SPI used as a wall material was developed to encapsulate *Enterococcus faecalis* HZNU P2 (*E. faecalis* HZNU P2) [76]. The results showed that encapsulation could offer sufficient protection to *E. faecalis* HZNU P2. The viability of encapsulated *E. faecalis* HZNU P2 did not decrease in SGF at pH 2.5 or 2.0 after 2 h incubation, while free cells were reduced from 11 log CFU/mL to 9.85 log CFU/mL in simulated gastric fluid (SGF, pH 2.5) at the same exposure time (Figure 9.1a,b). Only minor viability of encapsulated *E. faecalis* HZNU P2 lost in 1.0% or 2.0% bile salt solution for 1 or 2h

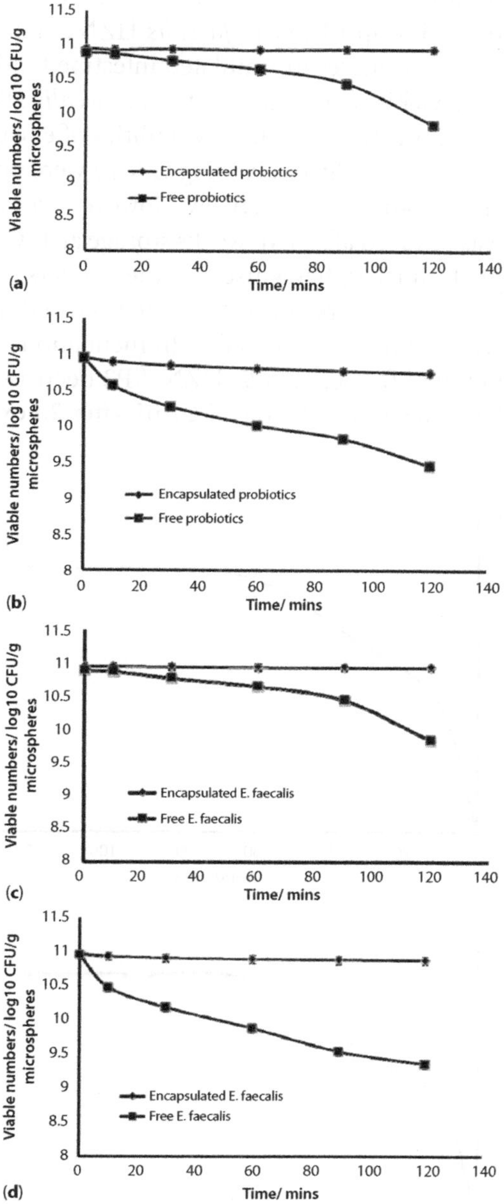

Figure 9.1 pH stability of free and encapsulated *E. faecalis* HZNU P2 in simulated gastric fluid (SGF) pH 2.5 and 2.0. (a)The stability of free and encapsulated *E. faecalis* HZNU P2 (SPI-alginate) in SGF pH 2.5. (b) The stability of free and encapsulated *E. faecalis* HZNU P2 (SPI-alginate) in SGF pH 2.0. (c) The stability of free and encapsulated *E. faecalis* HZNU P2 (soy milk-alginate) in SGF pH 2.5. (d) The stability of free and encapsulated *E. faecalis* HZNU P2 (soy milk-alginate) in SGF pH 2.0.

exposure, compared with no survival of free *E. faecalis* HZNU P2 under the same conditions. Encapsulated *E. faecalis* HZNU P2 was completely released from the microspheres in simulated intestine fluid (SIF) within 1 h (Figure 9.2a). The viability of encapsulated *E. faecalis* HZNU P2 stored for two weeks at 4 °C was fully retained. Viabilities of encapsulated *E. faecalis* HZNU P2, 9.6 Log CFU/g and 9.0 Log CFU/g were obtained at 25 °C and 37 °C after 21 days storage, respectively. However, around 1.0 log CFU/mL of free cells was reduced after two weeks storage at 4 °C (Figure 9.3a,b).

Soy milk–alginate microspheres used as the carriers were also developed in our group][77). The results so obtained showed that the viability of encapsulated cell did not decrease after 2h incubation in SGF at pH2.5 and 2.0. The viability of free *E. faecalis* HZNU P2 decreased from around 11 Log CFU/mL to 9.85 and 9.35 Log CFU/mL after 2h exporsure in SGF

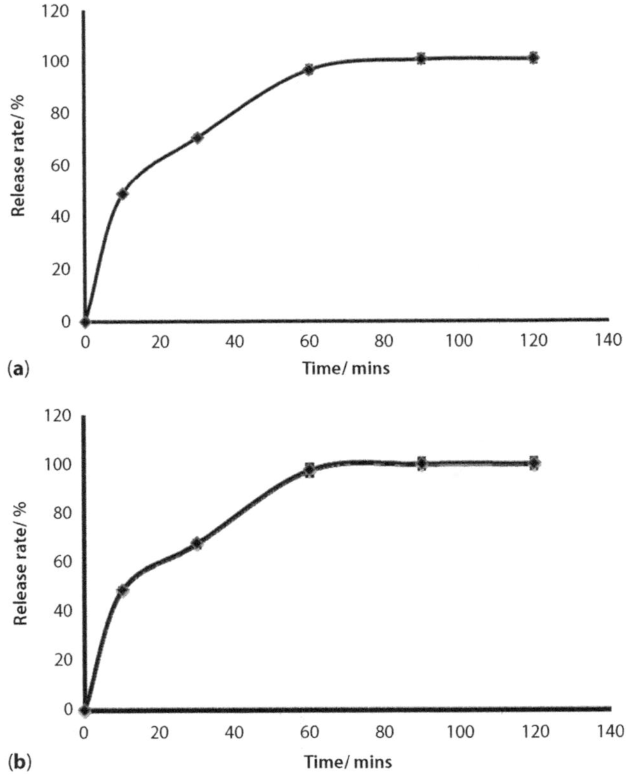

Figure 9.2 Release of encapsulated *E. faecalis* HZNU P2 in simulated intestine fluid (SIF). (a) SPI-Alginate microspheres. (b) Soy milk-Alginate microspheres.

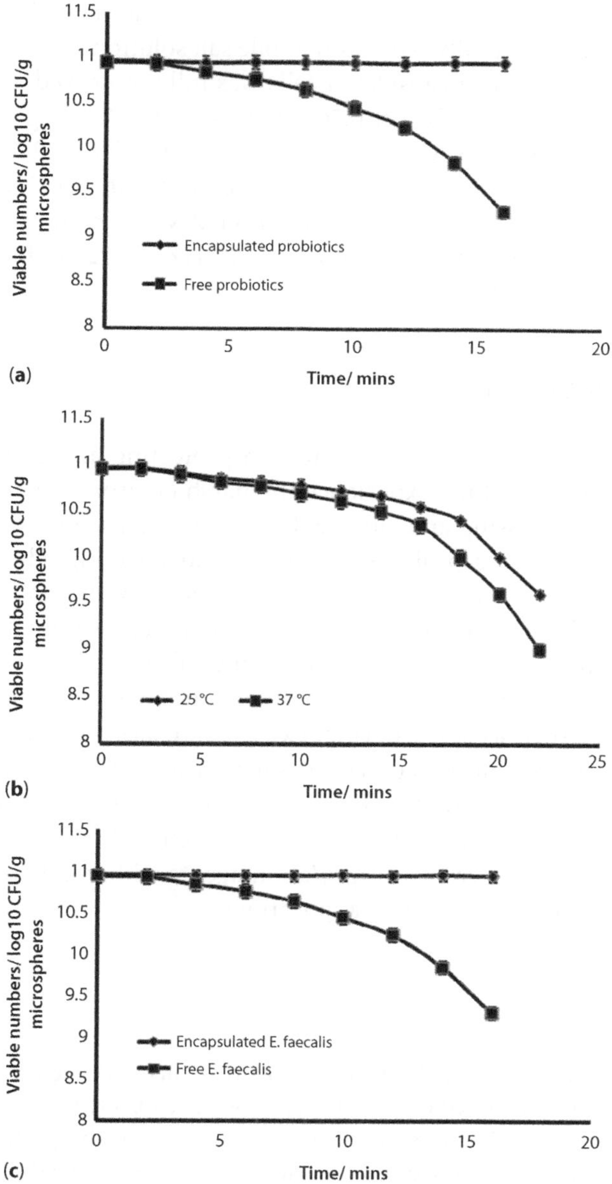

Figure 9.3 Storage stability of free and encapsulated *E. faecalis* HZNU P2 at 4, 25 °C and 37 °C. (a) Storage stability of free and encapsulated *E. faecalis* HZNU P2 at 4 °C (SPI-alginate). **(b)** Storage stability of encapsulated *E. faecalis* HZNU P2 at 25 and 37 °C (SPI-alginate). (c) Storage stability of free and encapsulated *E. faecalis* HZNU P2 at 4 °C (soy milk-alginate). (d) Storage stability of free and encapsulated *E. faecalis* HZNU P2 at 4 °C (soy milk-alginate).

at pH2.5 and 2.0, respectively (Figure 9.1c,d). Encapsulated cells experienced only minor viability reduction in bile salt solution after 1 or 2h treatment. The viability of encapsulated cells was fully preserved after 21 days of storage at 4°C, and only lost 0.70 Log CFU/g and 2.0 Log CFU/g after 21 days of storage at 25 °C and 37°C, respectively (Figure 9.3c,d). Figure 9.2b shows the release profile of encapsulated *E. faecalis* HZNU P2 in SIF, at 37°C over 2 h. All encapsulated *E. faecalis* HZNU P2 could be released in 60 min. Around 50% encapsulated *E. faecalis* HZNU P2 could be released after 10 min incubation.

9.4 Conclusion

Encapsulation plays an important role in protecting and delivering bioactive ingredients, which experience the harsh environmental conditions during food processing and storage. Currently, the use of soy proteins as wall materials for encapsulation of various bioactive ingredients, reflects the actual "green" tendency in the food industry. Although soy proteins have low solubility and poor functionality compared to animal-based proteins, it can be overcome through protein modification. Many studies have showed that different forms of soy proteins can be made using both various methods such as spray drying, extrusion, coacervate, and emulsion. Compared to other bio-based materials for encapsulation, such as polysaccharides or animal-derived proteins, soy proteins represent a very promising source of polymers with interesting functional properties, and can be produced into different encapsulation agents with good encapsulation efficiency and various potential applications.

References

1. Augustin, M.A., Hemar, Y., Nano-and micro-structured assemblies for encapsulation of food ingredients. *Chem. Soc. Rev.*, 38, 902–912, 2009.
2. Nesterenko, A., Alric, I., Silvestre, F., Durrieu, V., Vegetable proteins in microencapsulation: a review of recent interventions and their effectiveness. *Ind. Crops Prod.*, 42, 469–479, 2013.
3. Wan, Z.L., Guo, J., Yang, X.Q., Plant protein-based delivery systems for bioactive ingredients in foods. *Food Funct.*, 6, 2876–2889, 2015.
4. Chen, L., Remondetto, G.E., Subirade, M., Food protein-based materials as nutraceutical delivery systems. *Trends Food Sci. Tech.*, 17, 272–283, 2006.
5. Joye, I.J., Davidov-Pardo, G., McClements, D.J., Nanotechnology for increased micronutrient bioavailability. *Trends Food Sci. Technol.*, 40, 168–182, 2014.

6. Nesterenko, A., Alric, I., Silvestre, F., Durrieu, V., Influence of soy protein's structural modifications on their microencapsulation properties: α-tocopherol microparticles preparation. *Food Res. Int.*, 48, 387–396, 2012.
7. Benshitrit, R.C., Levi, C.S., Tal, S.L., Shimoni, E., Lesmes, U., Development of oral food-grade delivery systems: current knowledge and future challenges. *Food Funct.*, 3, 10–21, 2012.
8. Champagne, C.P., Fustier, P., Microencapsulation for the improved delivery of bioactive compounds into foods. *Curr. Opin. Biotech.*, 18, 184–190, 2007.
9. Velikov, K.P., Pelan, E., Colloidal delivery systems for micronutrients and nutraceuticals. *Soft Matter*, 4, 1964–1980, 2008.
10. McClements, D.J., Decker, E.A., Weiss, J., Emulsion-based delivery systems for lipophilic bioactive components. *J. Food Sci.*, 72, R109–R124, 2007.
11. McClements, D.J., Decker, E.A., Park, Y., Weiss, J., Structural design principles for delivery of bioactive components in nutraceuticals and functional foods. *Crit. Rev. Food Sci. Nutr.*, 49, 577–606, 2009.
12. Ubbink, J., Krüger, J., Physical approaches for the delivery of active ingredients in foods. *Trends Food Sci. Technol.*, 17, 244–254, 2006.
13. Gibbs, B.F., Kermasha, S., Alli, I., Mulligan, C.N., Encapsulation in the food industry: a review. *Int. J. Food Sci. Nutr.*, 50, 213–224, 1999.
14. Kamyshny, A., Magdassi, S., Microencapsulation, In: *Encyclopedia of Surface and Colloid Science*, P. Somasundaran, ed., Taylor & Francis, CRC Press, Boca Raton, FL, USA, pp. 3957–3969, 2006.
15. Foegeding, E.A., Davis, J.P., Food protein functionality: a comprehensive approach. *Food Hydrocolloid.*, 25, 1853–1864, 2011.
16. Lam, R.S., Nickerson, M.T., Food proteins: a review on their emulsifying properties using a structure-function approach. *Food Chem.*, 141, 975–984, 2013.
17. Jenkins, J.A., Breiteneder, H., Mills, E.N., Evolutionary distance from human homologs reflects allergenicity of animal food proteins. *J. Allergy Clin. Immunol.*, 120, 1399–1405, 2007.
18. Li, H., Zhu, K., Zhou, H., Peng, W., Effects of high hydrostatic pressure treatment on allergenicity and structural properties of soybean protein isolate for infant formula. *Food Chem.*, 132, 808–814, 2012.
19. Bujnowski, D., Xun, P., Daviglus, M.L., van Horn, L., He, K., Stamler, J., Longitudinal association between animal and vegetable protein intake and obesity among men in the United States: the Chicago western electric study. *J. Am. Diet. Assoc.*, 111, 1150–1155, 2011.
20. Sellmeyer, D.E., Stone, K.L., Sebastian, A., Cummings, S.R., A high ratio of dietary animal to vegetable protein increases the rate of bone loss and the risk of fracture in postmenopausal women. *Am. J. Clin. Nutr.*, 73, 118–122, 2001.
21. Rui-Henestrosa, V.P., Sanchez, C.C., Escobar, M.M.Y., Jimenez, J.J.P., Rodriguez, F.M., Patino, J.M.R., Interfacial and foaming characteristics of soy globulins as a function of pH and ionic strength. *Colloids Surf. B Biointerfaces*, 309, 202–215, 2007.

22. Gu, X., Campbell, L.J., Euston, S.R., Effects of different oils on the properties of soy protein isolate emulsions and gels. *Food Res. Int.*, 42, 925–932, 2009.
23. Reddy, N., Yang, Y., Potential of plant proteins for medical application. *Trends Biotechnol.*, 29, 490–498, 2011.
24. Dong, Q.Y., Chen, M.Y., Xin, Y., Qin, X.Y., Cheng, Z., Shi, L.E., Tang, Z.X., Alginate-based and protein-based materials for probiotics encapsulation: a review. *Int. J. Food Sci. Tech.*, 48, 1339–1351, 2013.
25. Dias, M.I., Ferreira, I.C.F.R., Barreiro, M.F., Microencapsulation of bioactives for food applications. *Food Funct.*, 6, 1035–1052, 2015.
26. de Kruif, C.G., Weinbreck, F., de Vries, R., Complex coacervation of proteins and anionic polysaccharides. *Curr.Opin. Colloid. In. Sci.*, 9, 340–349, 2004.
27. Tolstoguzov, V., Some thermodynamic considerations in food formulation. *Food Hydrocolloid.*, 17, 1–23, 2003.
28. Turgeon, S.L., Schmidt, C., Sanchez, C., Protein-polysaccaride complexes and coacervates. *Curr. Opin. Colloid In.*, 12, 166–178, 2007.
29. Lazko, J., Popineau, Y., Legrand, J., Soy glycinin microcapsules by simple coacervation method. *Colloids Surf B Biointerfaces*, 37, 1–8, 2004.
30. Leung, H.W., Ecotoxicology of glutaraldehyde: review of environmental fate and effects studies. *Ecotoxicol. Environ. Saf.*, 49, 26–39, 2001.
31. Strauss, G., Gibson, S.M., Plant phenolics as cross-linkers of gelatin gels and gelatin-based coacervates for use as food ingredients. *Food Hydrocolloid.*, 18, 81–89, 2004.
32. Dickinson, E., Food emulsions and foams: Stabilization by particles. *Curr. Opin. Colloid In.*, 15, 40–49, 2010.
33. Dickinson, E., Use of nanoparticles and microparticles in the formation and stabilization of food emulsions. *Trends Food Sci. Technol.*, 24, 4–12, 2012.
34. Can Karaca, A., Low, N., Nickerson, M., Emulsifying properties of canola and flaxseed protein isolates produced by isoelectric precipitation and salt extraction. *Food Res. Int.*, 44, 2991–2998, 2011.
35. Can Karaca, A, Low, N., Nickerson, M., Emulsifying properties of chickpea, faba bean, lentil and pea proteins produced by isoelectric precipitation and salt extraction. *Food Res. Int.*, 44, 2742–2750, 2011.
36. Ezpeleta, I., Irache, J.M., Stainmesse, S., Chabenat, C., Gueguen, J., Popineau, Y., Orecchioni, A.M., Gliadin nanoparticles for the controlled release of all-trans-retinoic acid. *Int. J. Pharm.*, 131, 191–200, 1996.
37. Lazko, J., Popineau, Y., Renard, D., Legrand, J., Microcapsules based on glycinin-sodium dodecyl sulfate complex coacervation. *J. Microencapsul.*, 21, 59–70, 2004.
38. Jiang, J., Chen, Xiong, Y., Structural and emulsifying properties of soy protein isolate subjected to acid and alkaline pH-shifting processes. *J. Agric. Food Chem.*, 57, 7576–7583, 2009.
39. Augustin, M.A., Sanguansri, L., Bode, O., Maillard reaction products as encapsulants for fish oil powders. *J. Food Sci.*, 71, E25–E32, 2006.

40. Jourdain, L., Leser, M.E., Schmitt, C., Michel, M., Dickinson, E., Stability of emulsions containing sodium caseinate and dextran sulfate: relationship to complexation in solution. *Food Hydrocolloid.*, 22, 647–659, 2008.
41. Guo, J., Yang, X.Q., He, X.T., Wu, N.N., Wang, J.M., Gu, W., Zhang, Y.Y., Limited aggregation behavior of β-conglycinin and its terminating effect on glycinin aggregation during heating at pH 7.0. *J. Agric. Food Chem.*, 60, 3782–3791, 2012.
42. Chen, L., Subirade, M., Elaboration and characterization of soy/zein protein microspheres for controlled nutraceutical delivery. *Biomacromolecules*, 10, 3327–3334, 2009.
43. Chen, L., Hébrard, G., Beyssac, E., Denis, S., Subirade, M., In vitro study of the release properties of soy–zein protein microspheres with a dynamic artificial digestive system. *J. Agric. Food Chem.*, 58, 9861–9867, 2010.
44. Rascon, M.P., Beristain, C.I., Garcie, H.S., Salgado, M.A., Carotenoid retention and storage stability of spray-dried paprika oleoresin using gum Arabic and soy protein isolate as wall materials. *Food Sci. Technol.*, 44, 549–557, 2011.
45. Tang, C.H., Li, X.R., Microencapsulation properties of soy protein isolate and storage stability of the correspondingly spray-dried emulsions. *Food Res. Int.*, 52, 419–428, 2013.
46. Zhang, J., Liang, L., Tian, Z., Chen, L., Subirade, M., Preparation and in vitro evaluation of calcium-induced soy protein isolate nanoparticles and their formation mechanism study. *Food Chem.*, 133, 390–399, 2012.
47. Teng, Z., Luo, Y., Wang, T., Zhang, B., Wang, Q., Development and application of nanoparticles synthesized with folic acid conjugated soy protein. *J. Agric. Food Chem.*, 61, 2556–2564, 2013.
48. Zhang, J., Field, C.J., Vine, D., Chen, L., Intestinal uptake and transport of vitamin B_{12}-loaded soy protein nanoparticles. *Pharm. Res.*, 32, 1288–1303, 2015.
49. Teng, Z., Luo, Y., Wang, Q., Carboxymethyl chitosan–soy protein complex nanoparticles for the encapsulation and controlled release of vitamin D_3. *Food Chem.*, 141, 524–532, 2013.
50. Ding, X., Yao, P., Soy protein/soy polysaccharide complex nanogels: folic acid loading, protection, and controlled delivery. *Langmuir*, 29, 8636–8644, 2013.
51. Gao, Z.M., Wang, J.M., Wu, N.N., Wan, Z.L., Guo, J., Yang, X.Q., Yin, S.W., Formation of complex interface and stability of oil-in-water (O/W) emulsion prepared by soy lipophilic protein nanoparticles. *J. Agric. Food Chem.*, 61, 7838–7847, 2013.
52. Gao, Z.M., Zhu, L.P., Yang, X.Q., He, X.T., Wang, J.M., Guo, J., Qi, J.R., Wang, L.J., Yin, S.W., Soy lipophilic protein nanoparticles as a novel delivery vehicle for conjugated linoleic acid. *Food Funct.*, 5, 1286–1293, 2014.
53. Nishinari, K., Fang, Y., Guo, S., Phillips, G., Soy proteins: a review on composition, aggregation and emulsification. *Food Hydrocolloid.*, 39, 301–318, 2014.

54. Djordjevic, D., McClements, D.J., Decker, E.A., Oxidative stability of whey protein-stabilized oil-in-water emulsions at pH 3: potential ω-3 fatty acid delivery systems. *J. Food Sci.*, 69, C356–C362, 2006.
55. Tapal, A., Tiku, P.K., Complexation of curcumin with soy protein isolate and its implications on solubility and stability of curcumin. *Food Chem.*, 130, 960–965, 2012.
56. Tang, C.H., Liu, F., Soy protein nanoparticle aggregates as pickering stabilizers for oil-in-water emulsions. *J. Agric. Food Chem.*, 61, 8888–8898, 2013.
57. Wan, Z.L., Wang, J.M., Wang, L.Y., Yang, X.Q., Yuan, Y., Enhanced physical and oxidative stabilities of soy protein-based emulsions by incorporation of a water-soluble stevioside–resveratrol complex. *J. Agric. Food Chem.*, 61, 4433–4440, 2013.
58. Yuan, Y., Wan, Z.L., Yin, S.W., Yang, X.Q., Stability and antimicrobial property of soy protein/chitosan mixed emulsion at acidic condition. *Food Funct.*, 4, 1394–1401, 2013.
59. Wan, Z.L., Wang, J.M., Wang, L.Y., Yuan, Y., Yang, X.Q., Complexation of resveratrol with soy protein and its improvement on oxidative stability of corn oil/water emulsions. *Food Chem.*, 161, 324–331, 2014.
60. Jun-xia, X., Hai-yan, Y., Jian, Y., Microencapsulation of sweet orange oil by complex coacervation with soybean protein isolate/gum Arabic. *Food Chem.*, 125, 1267–1272, 2011.
61. Mendanha, D.V., Ortiz, S.E.M., Favaro-Trindade, C.S., Mauri, A., Monterrey-Quintero, E.S., Thomazini, M., Microencapsulation of casein hydrolysate by complex coacervation with SPI/pectin. *Food Res. Int.*, 42, 1099–1104, 2009.
62. Rusli, J.K., Sanguansri, L., Augustin, M.A., Stablization of oils by microencapsulation with heated protein-glucose syrup mixtures. *J. Am. Oil Chem. Soc.*, 83, 965–971, 2006.
63. Vega-Lugo, A.C., Lim, L.T., Controlled release of allyl isothiocyanate using soy protein and poly (lactic acid) electrospun fibers. *Food Res. Int.*, 42, 933–940, 2009.
64. Wang, S., Marcone, M.F., Barbut, S., Lim, L.T., Electrospun soy protein isolate-based fiber fortified with anthocyanin-rich red raspberry (Rubus strigosus) extracts. *Food Res. Int.*, 52, 467–472, 2013.
65. Hoffman, A.S., Hydrogels for biomedical applications. *Adv. Drug Deliv. Rev.*, 54, 3–12, 2002.
66. Kopeček, J., Hydrogel biomaterials: a smart future? *Biomaterials*, 28, 5185–5192, 2007.
67. Peppas, N.A., Hydrogels and drug delivery. *Curr. Opin. Colloid In.*, 2, 531–537, 1997.
68. Maltais, A., Remondetto, G.E., Subirade, M., Soy protein cold-set hydrogels as controlled delivery devices for nutraceutical compounds. *Food Hydrocolloid.*, 23, 1647–1653, 2009.
69. Maltais, A., Remondetto, G.E., Subirade, M., Tabletted soy protein cold-set hydrogels as carriers of nutraceutical substances. *Food Hydrocolloid.*, 24, 518–524, 2010.

70. Caillard, R., Remondetto, G., Subirade, M., Physicochemical properties and microstructure of soy protein hydrogels co-induced by Maillard type cross-linking and salts. *Food Res. Int.*, 42, 98–106, 2009.
71. Caillard, R., Mateescu, M., Subirade, M., Maillard-type cross-linked soy protein hydrogels as devices for the release of ionic compounds: an in vitro study. *Food Res. Int.*, 43, 2349–2355, 2010.
72. Guo, J., Zhang, Y., Yang, X.Q., A novel enzyme cross-linked gelation method for preparing food globular protein-based transparent hydrogel. *Food Hydrocolloid.*, 26, 277–285, 2012.
73. Favaro-Trindade, C.S., Santana, A.S., Monterrey-Quintero, E.S., Trindade, M.A., Netto, F.M., The use of spray drying technology to reduce bitter taste of casein hydrolysate. *Food Hydrocolloid.*, 24, 336–340, 2010.
74. Ortiz, S.E.M., Mauri, A., Monterrey-Quintero, E.S., Trindade, M.A., Production and properties of casein hydrolysate microencapsulated by spray drying with soybean protein isolate. *Food Sci. Technol.*, 42, 919–923, 2009.
75. de Conto, L.C., Grosso, C.R.F., Goncalves, L.A.G., Chemometry as applied to the production of omega-3 microcapsules by complex coacervation with soy protein isolate and gum Arabic. *LWT-Food Sci. Technol.*, 53, 218–224, 2013.
76. Zhang, Y., Zheng, W., Gu, J.F., Ni, J., Wang, L., Tang, Z.X., Shi, L.E., Soy protein isolate-alginate microspheres for encapsulation of *Enterococcus faecalis* HZNU P2. *Braz. Arch. Biol. Technol.*, 58, 805–811, 2015.
77. Shi, L.E., Zheng, W., Zhang, Y., Liu, X.Y., Tang, Z.X., Soy milk-based microspheres as potential carriers for the protection of *Enterococcus faecalis* HZNU P2. *J. Funct. Foods*, 18, 487–491, 2015.
78. Yu, C., Wang, W., Yao, H., Liu, H., Preparation of phospholipid microcapsules by spray drying. *Dry Technol.*, 25, 695–702, 2007.
79. Charve, J., Reineccius, G.A., Encapsulation performance of proteins and traditional materials for spray dried flavors. *J. Agric. Food Chem.*, 57, 2486–2492, 2009.
80. Nori, M.P., Favaro-Trindade, C.S., de Alencar, S.M., Thomazini, M., de Camargo Balieiro, J.C., Castillo, C.J.C., Microencapsulation of propolis extract by complex coacervation. *LWT-Food Sci. Technol.*, 44, 429–435, 2011.

Index

Also of Interest from Scrivener Publishing by Visakh P.M

Nanostructured Polymer Membranes
Volume 1: Processing and Characterization
Edited by Visakh P. M and Olga Nazarenko
Published 2016. ISBN: 978-1-118-83173-1

Nanostructured Polymer Membranes
Volume 2: Applications
Edited by Visakh P. M and Olga Nazarenko
Published 2016. ISBN 978-1-118-83178-6

Polyethylene-based Blends, Composites and Nanocomposites
Edited by Visakh P. M and María José Martínez Morlanes
Published 2015. ISBN 978-1-118-83128-1

Polyoxymethylene Handbook
Structure, Properties, Applications and Their Nanocomposites
Edited by Sigrid Lüftl, Visakh P.M., and Sarath Chandran
Published 2014. ISBN: 978-1-118-38511-1

Advances in Food Science and Technology
In two volumes
Edited by Visakh, P.M., Sabu Thomas, Laura B. Iturriaga, and Pablo Daniel Ribotta
Published 2013. 978-1-118-12102-3

Handbook of Engineering and Specialty Thermoplastics 4
Volume 4: Nylons
Edited by Sabu Thomas and Visakh P.M
Published 2012. ISBN 978-0-470-63925-2

Handbook of Engineering and Specialty Thermoplastics 3
Volume 3: Polyethers and Polyesters
Edited by Sabu Thomas and Visakh P.M.
Published 2012. ISBN 978-0-470-63926-9

Printed and bound by CPI Group (UK) Ltd, Croydon, CR0 4YY
17/09/2026
14964241-0002